Springer Series in **Materials Science** 23

Edited by Ulrich Gonser

1 **Chemical Processing with Lasers**
By D. Bäuerle

2 **Laser-Beam Interactions with Materials**
Physical Principles and Applications
By. M. von Allmen

3 **Laser Processing of Thin Films and Microstructures**
Oxidation, Deposition and Etching of Insulators
By. I. W. Boyd

4 **Microclusters**
Editors: S. Sugano, Y. Nishina, and S. Ohnishi

5 **Graphite Fibers and Filaments**
By M. S. Dresselhaus, G. Dresselhaus, K. Sugihara, I. L. Spain, and H. A. Goldberg

6 **Elemental and Molecular Clusters**
Editors: G. Benedek, T. P. Martin, and G. Pacchioni

7 **Molecular Beam Epitaxy**
Fundamentals and Current Status
By M. A. Herman and H. Sitter

8 **Physical Chemistry of, in and on Silicon**
By G. F. Cerofolini and L. Meda

9 **Tritium and Helium-3 in Metals**
By R. Lässer

10 **Computer Simulation of Ion-Solid Interactions**
By W. Eckstein

11 **Mechanisms of High Temperature Superconductivity**
Editors: H. Kamimura and A. Oshiyama

12 **Dislocation Dynamics and Plasticity**
By T. Suzuki, S. Takeuchi, and H. Yoshinaga

13 **Semiconductor Silicon**
Materials Science and Technology
Editors: G. Harbecke and M. J. Schulz

14 **Graphite Intercalation Compounds I**
Structure and Dynamics
Editors: H. Zabel and S. A. Solin

15 **Crystal Chemistry of High-$T_c$ Superconducting Copper Oxides**
By B. Raveau, C. Michel, M. Hervieu, and D. Groult

16 **Hydrogen in Semiconductors**
By S. J. Pearton, M. Stavola, and J. W. Corbett

17 **Ordering at Surfaces and Interfaces**
Editors: A. Yoshimori, T. Shinjo, and H. Watanabe

18 **Graphite Intercalation Compounds II**
Editors: S. A. Solin and H. Zabel

19 **Laser-Assisted Microtechnology**
By S. M. Metev and V. P. Veiko

20 **Microcluster Physics**
By S. Sugano

21 **The Metal-Hydrogen System**
By Y. Fukai

22 **Ion Implantation in Diamond, Graphite and Related Materials**
By M. S. Dresselhaus and R. Kalish

23 **The Real Structure of High-$T_c$ Superconductors**
Editor: V. Sh. Shekhtman

24 **Metal Impurities in Silicon Device Fabrication**
By K. Graff

25 **Optical Properties of Metal Clusters**
By U. Kreibig and M. Vollmer

26 **Gas Source Molecular Beam Epitaxy**
Growth and Properties of Phosphorus Containing III-V Heterostructures
By M. B. Panish and H. Temkin

V. Sh. Shekhtman (Ed.)

# The Real Structure of High-$T_c$ Superconductors

With 107 Figures

Springer-Verlag
Berlin Heidelberg New York
London Paris Tokyo
Hong Kong Barcelona
Budapest

Professor Dr. Veniamin Sh. Shekhtman

Institute for Solid State Physics
Russian Academy of Sciences
Moscow District
Chernogolovky 142432, Russia, CIS

(All contributors are at the same address.)



ISBN 3-540-56559-0 Springer-Verlag Berlin Heidelberg New York
ISBN 0-387-56559-0 Springer-Verlag New York Berlin Heidelberg

Library of Congress Cataloging-in-Publication Data. The Real structure of high-Tc superconductors / Veniamin Sh. Shekhtman. p. cm. – (Springer series in materials science ; v. 23) Includes bibliographical references and index. ISBN 3-540-56559-0 (Berlin: alk. paper). – ISBN 0-387-56559-0 (New York: alk. paper) 1. High temperature superconductors–Structure. 2. Crystals–Structure. I. Shekhtman, Veniamin Sh., 1929- . II. Series.
QC611.98.H54R4 1993 537.6'231–dc20 93-13429



Typesetting: PS™ Technical Word Processor
54/3140 - 5 4 3 2 1 0 - Printed on acid-free paper

# Preface

High-Temperature SuperConductivity (HTSC) was first observed less than six years ago. However, the number of publications on this topic is, perhaps, commensurate with the total number of projects carried out since the Kammerling-Onnes discovery. At present, the experimental observation of a transition temperature near 100 K or even somewhat higher does not seem to be an extraordinary achievement. There is also progress in striving for the critical current parameters and magnetic-field strengths, which do support optimistic predictions. Nevertheless, further concentrated efforts are necessary. The level of understanding of these phenomena is still rather modest. The study of complex oxides, which served as a basis for this vital breakthrough in superconductivity, is a broad area in which physicists and physical chemists using various methods can interact.

Aspects of the real (in contrast to the ideal) structure of these compounds are discussed in terms of their effects on the properties of HTSC materials. The role of boundaries and of substructural elements in determining the critical parameters of superconductivity must be understood in order to be able to design and construct practical superconducting devices. In this quasi-monograph we combine the results of analysis of microstructures using diffraction and microscopy with the detailed investigations of the magnetic configuration in a mixed state. This provides a more complete and self-consistent picture of the internal composition of superconducting oxides.

Special attention is given to the "oxygen problem", that is, to the unexpected features which can be traced to "oxygen impurity atoms", the ordering, and possible isomorphous substitutions in structures similar to perovskites. Deformation and the mechanical influence on the structure and properties of HTSC are also discussed.

The work was performed in the framework of the Russian State Research and Technology Program 'High-Temperature Superconductivity'.

Chernogolovka, Moscow District
December 1992

*V.Sh. Shekhtman*

# Contents

# 1. Introduction

V.Sh. Shekhtman

The notion of a real crystal was developed at the beginning of the century. Now it is commonly accepted that it is impossible to describe fundamental phenomena of crystalline bodies without resorting to the physical picture of an ideal (harmonic) atomic lattice with "defects". The quotation marks attached to the word "defects" since we are convinced that deviations from a regular structure - namely grains, blocks, boundaries, point defects, etc. - should be considered as legitimate components of crystalline materials. The many impressive demonstrations of the fact that the system of defects defines the distinguishing properties of a given material attest to this statement. This is the basis for all semiconductor physics, the physics of strength and plasticity, magnetic materials, etc.

High-Temperature SuperConductors (HTSC) should not be an exception in this respect. The necessity for more detailed study of the atomic processes occurring in the substructure during HTSC-crystal growth has become ever more obvious. The peculiarities of these materials lead to many new experimental obstacles not known in steel or other alloys. The complexity is already seen in the initial stages of investigation: for the $YBa_2Cu_3O_{7-\delta}$ system (frequently abbreviated by $1\text{-}2\text{-}3\text{-}O_{7-\delta}$ with 1-2-3 being a shorthand for the number of respective atoms) the crystal appears to be a product of an uncontrolled martensite transformation which splits the medium into many structural domains of a superconducting phase. Compositions based on Bi or Tl are characterized by structural modulations of shear and ordering as well as unusual cell sizes. LaSr materials, on the other hand, can be described as disordered solid solutions. All these peculiarities require serious attention. In comparison, the structure of alloys of iron, aluminum and titanium, or A-15-type compounds, the investigation of which involved the tremendous efforts of G. Kurdyumov and A. Guinier and other leading figures of classical metallurgy, present a different picture. In particular, it refers to the role of oxygen. Addition of the nonmetal is the determining factor in the properties of new materials. From a chemical point of view, the role played by oxygen atoms differs, for example, from that of carbon in a cell of iron or silicon in $V_3Si$. Accordingly, in HTSC crystals the ordering and valence states of metal atoms, and the selective mobility of crystallographically different oxygen atoms are only some of the interesting but experimentally difficult problems that remain to be studied.

A remark on the logic of scientific investigation is appropriate before continuing. *Max von Laue* noted that often the discovery of a new pheno-

Springer Series in Materials Science, Vol. 23
**The Real Structure of High-$T_c$ Superconductors**
Editor: V.Sh. Shekhtman 

menon suddenly joins previously separated spheres of science. In his book on the history of physics [1.1], he provided examples: the joining of optics with thermodynamics, and diffraction theory with crystallography. We see another example of this in a new chapter of the Curriculum Vitae Physica: the new superconductivity requires that we refocus our attention from substances of the "old metal world", i.e. solid solutions and intermetallics, to a "green ferroelectric branch". The family of perovskite crystals has begun to play a new role in the physics of superconducting materials. It seems strange now that ten to fifteen years ago this editor could hardly anticipate the geometric similarity of twinning under phase transitions in metallic $V_3Si$ and ferroelectric $BaTiO_3$ [1.2, 3]. The progress in HTSC investigations has demonstrated, however, that this analogy is not just a formal one.

## 1.1 About This Book

The contributors to this quasi-monograph have combined their efforts to summarize the results of investigations on HTSC crystals performed at the Institute of Solid State Physics of the Russian Academy of Sciences, Chernogolovka.

Chapter 2 presents an analysis of data on dislocation configurations obtained by high-resolution electron microscopy and electron diffraction. The pattern in electron micrographs, indicating transitional zones, related, e.g., to a violated sequence of $CuO_2$ layers, are emphasized. Independent results of a study of twin structures in yttrium-based materials are presented. The features of electron-microscopy observations of a phase transition "in situ" are discussed in detail. Twin-configuration changes due to repeated heating by an electron beam are studied, as well as variations in the phase composition of a rapidly-cooled material after heating in a microscope column.

The main emphasis of Chap.3 is on the polysynthetic structure which is formed in the martensite mechanism. Such structural transformations via homogeneous deformation of twinning, first established for iron and titanium alloys, appeared to be typical of many ferroelectric transitions, and now for HTSC crystals as well. In these transitions the crystal loses its translational periodicity but retains a "memory" of the initial axes. This means that the number of structurally equivalent orientation states and the configuration of twin boundaries are defined by the parent structure. In superconductors, a mutual fitting of domains on their boundaries appears to be significant for values of the critical parameters of current and magnetic field. Moreover, a smooth reorientation of neighboring domains can be achieved through the layers of a previous phase ("parent austenite", in metallurgical terms). In addition, 1-2-3 crystals require a reordering of oxygen atoms at the transition of domains from one system to another. The results collected in Chap.3 demonstrate that various associations of regions with different degrees of ordering of oxygen atoms can arise. Significantly extended regions, separating twins of different orientation, were discovered. In addition, quasi-twin states have been defined, in which an entire crystal is virtually just the set of transitional regions free of regular domains.

Chapter 4 is devoted to deformation characteristics. First, the dependence of the plasticity of $1\text{-}2\text{-}3\text{-}O_{7-\delta}$ ceramics on the oxygen parameter $\delta$ is considered. It has been found that both an oxygen deficiency ($\delta = 0.9 \div 0.8$) and saturation ($\delta = 0.1 \div 0.2$) lead to brittleness. Thus, the highest plasticity occurs at intermediate $\delta$ values.

This proves the active role of oxygen and oxygen holes in the mechanism of plastic deformation. We believe that shears, shifts of dislocations connected with the plastic deformation are accompanied (or induced) by displacements of oxygen-free positions, that is, by reordering. An unexpected role played by defects is manifested.

Another important aspect which is touched on in Chap.4 is the role of twin boundaries in the determination of critical parameters. A comparison is made to superconducting alloys; in particular, it is interesting to point out the specific properties of transformation twins in HTSC as distinguished, e.g., from mechanical twins in niobium alloys.

The interaction of a vortex lattice in a superconductor with elements of the real structure is treated in the two following chapters. Chapter 5 focuses on techniques for decorating the *Abrikosov vortices* by dispersed ferromagnetic powders and subsequent microscopy analysis. The values of the quantum of magnetic flux in HTSC, which permit the confirmation of the mechanism of carrier coupling, are measured. A major part of the work is the determination of vortex lattice features in 1-2-3 materials and Bi or Tl systems as a function of temperature and magnetic fields. New configurations of distorted vortex structure, including liquid-like violation of order, are established. Special attention is paid to the anisotropic properties under the influence of a tilted magnetic field. Experimental efforts are also directed to revealing the immediate effects of vortex interaction with a twin structure. Solving this problem, connected with the fundamental one of pinning, one manages to obtain the quantitative characteristics of the interaction potential and an estimation of the elementary pinning force acting on a unit of vortex length.

Polarization methods are the subject of Chap.6. These yield a view of the HTSC microstructure by magnetooptic iron-garnet films. The twin structure of 1-2-3 compounds, which was varied by a thermomechanical treatment, was investigated in particular. A novel technique, which uses ferrimagnetic films to indicate the distribution of a magnetic field, allows not only the observation of the magnetic flux but also the measurement of the local transition temperature at different points of a sample. Thus, quantitative estimates of critical fields and currents can be made, and their thermal variation and anisotropy determined. The correlation of magnetic properties with the twin structure of as-grown crystals was also determined. A more thorough investigation revealed that variations of oxygen content, effecting the twin structure, are primarily responsible for the observed changes of superconducting characteristics.

The data presented in Chap. 7, which is devoted to anion isomorphism, are directly related to the oxygen problem discussed above. Here we dwell on establishing the role of oxygen atoms in inducing superconductivity. The unique properties of the O1 position, occupied by oxygen in YBaCuO, permitted a search for the probable position of halogen atoms in a given regular system of points. Discussed are in detail the methods of halogenation of ceramics, films and single crystals. These experiments show that an initially tetragonal crystal (lacking oxygen) can be reactivated, that is, superconducting properties can be restored by treating it with halogen vapors in an oxygen-free atmosphere. Considerable experimental efforts are underway to show whether the halogen atoms enter into the lattice of an oxygen-deficient compound.

The use of Mössbauer spectroscopy is the subject of the last chapter (Chap. 8), and is discussed in application to the problem of cation substitution of copper atoms and the magnetic properties of an HTSC basic structure. Experimental results for the behavior of iron used as a doping component are presented. The parameters governing the occurrence of an antiferromagnetic order are found. The first data for the effects of various oxygen positions in a new 1-4-3 composition are obtained.

## References


1.1 M. von Laue: *Geschichte der Physik* (Athenaum, Bonn 1950) pp.1-15

1.2 V. Marchenko, L. Shabelnikov, V. Shekhtman: Fiz. Tverd. Tela **17**, 2883-2888 (1975)

1.3 S. Aknazarov, L. Shabelnikov, V. Shekhtman: Fiz. Tverd. Tela **17**, 30-34 (1975)

# 2. Electron-Microscopy Investigation of the Structure of Defects

V.A. Goncharov and E.V. Suvorov

Since that moment when the superconducting compounds of Y-Ba-Cu-O, Bi-Sr-Ca-Cu-O, Tl-Ba-Ca-Cu-O type were discovered, the quasi-two-dimensional structure of the superconducting phases has attracted much interest. In Y-Ba-Cu-O compounds there are twin boundaries of high bulk density, in Bi-Sr-Ca-Cu-O compounds there are structural modulations, and in the Tl-Ba-Ca-Cu-O compounds they exhibit polytypism. Up to the present time, a number of works have been devoted to the influence of these two-dimensional defects, as supposed pinning centers, on the superconducting properties and as special crystal planes where 2D and 1D models of superconductivity may be realized. Since the role of quasi-2D defects on the physical properties of superconducting compounds is not yet quite clear, the study of defect structures in these materials is of significant value. In particular, the study of oxygen atoms and oxygen vacancies inside and in the vicinity of planar defects presents an especially interesting problem, since, in the majority of existing models of high-temperature superconductivity, the localization of oxygen ions is one of the basic factors affecting superconductivity parameters.

## 2.1 The Structure of Defects in $YBa_2Cu_3O_{7-x}$

The structure of an orthorhombic phase, possessing the best superconducting properties of an ideal $YBa_2Cu_3O_7$ (named 1-2-3) composition, is described by alternating four layers of BaO, CuO and $Cu_2O$ in the direction of the $c$-axis: BaO-CuO-BaO-$CuO_2$-Y-$CuO_2$-BaO-CuO-BaO type. There is a 1D chain [-O-Cu-O-Cu-O-] in the direction of the $b$-axis in the CuO basal layer. Each atom of copper in the chain is surrounded by four oxygen atoms forming a planar square $CuO_4$. Each atom of copper in the $CuO_2$ layers is surrounded by five oxygen atoms forming a square pyramid. It is known, however, that there is always a deviation from ($YBa_2Cu_3O_{7-x}$) stoichiometry in real single crystals, which is due primarily to free oxygen sites in the CuO layers [2.1-3].

### 2.1.1 Planar Defects of Non-Twin Nature. Dislocations

High-Resolution Electron Microscopy (HREM) revealed planar defects of a non-twin nature, they seem to be caused by non-stochiometric oxygen de-

Springer Series in Materials Science, Vol. 23
**The Real Structure of High-$T_c$ Superconductors**
Editor: V.Sh. Shekhtman 

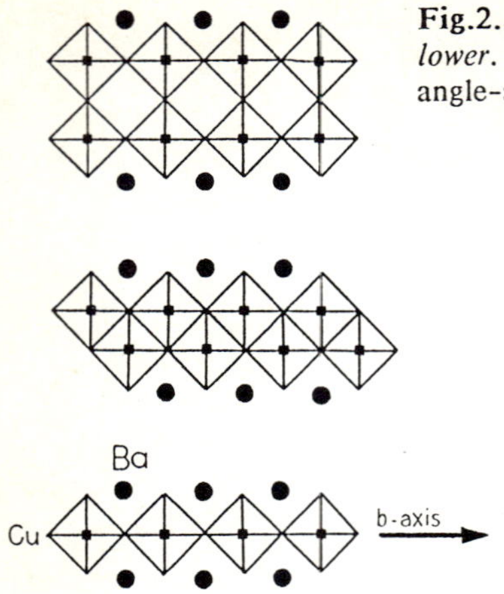


**Fig.2.1.** One-dimensional CuO chain projected in [100] *lower*. Two possible variants of conjugation of samples: angle-angle (*middle*) side-side (*upper*)

ficiency in $Ba_2YCu_3O_{6.9}$ single crystals [2.4]. Later, at least two types of such defects were discovered by HREM methods [2.1]. The observations were made in (010), (110) sections of single crystals under diffraction conditions where atoms of metals were seen as dark spots. The projections of plane defects on the observation plane appeared as white areas extended along a single axis, corresponding to increasing distances between barium metal layers. The spacing between two neighboring Ba-Ba layers was enlarged to 0.60÷0.65 nm, that is, to 0.2 nm more than the normal range. The proposed model of such a plane defect is described by two 1D CuO chains, arbitrarily inserted into basal plane, to form a double chain which extends in the *b*-direction. Figure 2.1 depicts the projection of a 1D CuO chain perpendicular to *a* and two possible ways of parallel connection of such chains. The "side-side" edgewise connection (middle picture) yields better agreement with the value of the Ba-Ba layer distance. Then the sequence of alternating layers along the *c*-axis in the presence of such a planar defect is

$$CuO_2\text{-}Y\text{-}CuO_2\text{-}BaO \boxed{\text{-}CuO\text{-}CuO\text{-}} BaO\text{-}CuO_2\text{-}Y\text{-}CuO_2$$

$$\downarrow$$

$$\boxed{\text{region of defect}}$$

Figure 2.2 exhibits a sketch of this defect. Another type of planar defect is observed in the (010) section. The bright white spots show the CuO basal layers separated from each other by 1.17 nm. "Relaxation" regions with the deflection vector $1/2a$ are also observed parallel to the defect plane. The defect may be interpreted as a local "swell" of the lattice in the *c*-direction.

The above can be represented as excess $CuO_2$ layers inserted near yttrium layers and forming double pyramids (Fig.2.3). The sequence of layers is then

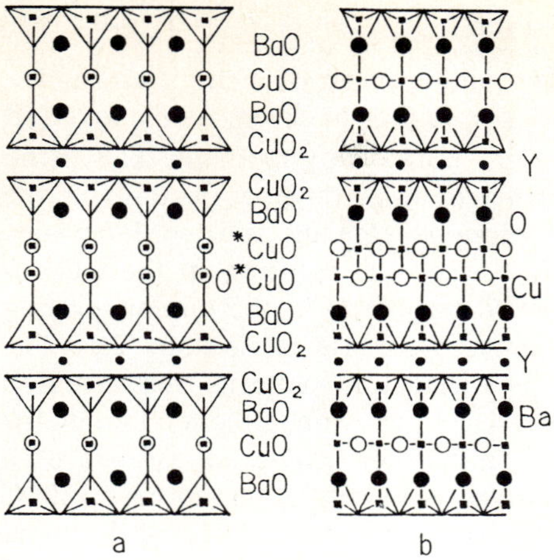


**Fig.2.2.** Model of a planar defect formed by two CuO chains with the "side-side" conjugation. Projections along the (*a*) [010] and (*b*) [100] directions

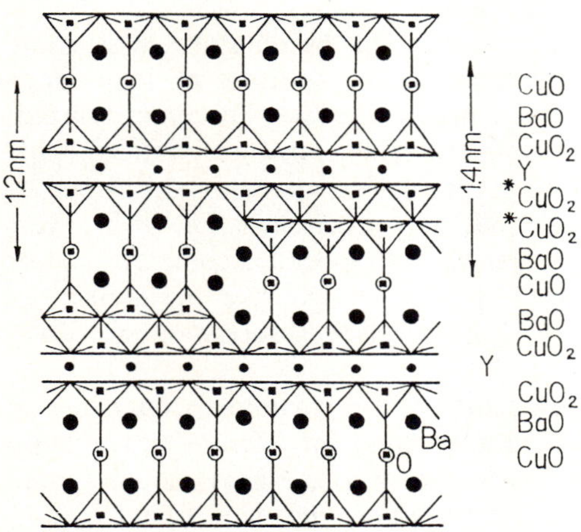


**Fig.2.3.** Model of planar defect formed by additional $CuO_2$ layers inserted near Y layers, forming double pyramids

BaO-CuO-BaO | $-CuO_2-CuO_2-$ | BaO-CuO-BaO

↓

region of defect

This type of planar defect may be considered as a prototype of an edge dislocation center and, under certain conditions, together with symmetric areas of relaxation (deflection vector $1/2a$), it can be seen as an edge dislocation with the Burgers vector $a/2$ [100]. According to our electron-microscopy observations and experiments connected with the contrast of growth dislocations and dislocations of introduced deformations, the dislocations with the Burgers vectors [100] and [010] dominate. The same data have been obtained by another group [2.5].

### 2.1.2 Twin Boundaries and Twin Layers

The most common defects in the orthorhombic phase of a single and polycrystals $YBa_2Cu_3O_{7-x}$ are twin boundaries and twin layers easily revealed in the (001) observation plane by electron-microscopy methods. The twinning planes are (110) and ($1\bar{1}0$). Splitting of long-range reflections in the Electron Diffraction Pattern (EDP) beginning from the third-order peaks, proves the existence of twin symmetry. Since the splitting in the EDP is small, laser-beam diffraction from small regions of a diffraction grating formed by a high-resolution negative image were also carried out [2.6]. The coincidence of the electron and optical diffraction patterns demonstrated that the small splitting in higher-order reflections is unambiguous proof for reflection symmetry in the (100) and (110) planes. The appearance of a large number of coherent twins of this type is readily observed near 750° C when the sample is slowly cooled from a higher temperature while being saturated with oxygen. A phase transition from a tetragonal state to an orthorhomic one takes place at the same temperature.

The orthophase-tetraphase transformation can be obtained in situ under "soft" heating by an electron beam [2.7]. The splitting of long-range -order spots disappears and the image of microtwin layers becomes blurred. It was possible to reversibly thermocycle the same transformation with repeated occurrence of a twin structure several times, though a detailed picture of the twin boundary localization was not reproduced. The twin layers became narrower, overlapped more often and disappeared completely after several cycles. *Van Tendeloo* et al. [2.7] believed that the mechanism of orthophase-tetraphase transformation is the ordering-disordering of vacancies in the oxygen sublattice.

It has been seen that the disordering of heavy atoms occurs due to irregular substitutions of perovskite cubic centers for barium and yttrium atoms [2.8]. Such highly disordered (with respect to the heavy atoms) materials were obtained after rapid cooling and annealing of a $YBa_2Cu_3O_{7-x}$ compound during several hours at a temperature lower than that required to form ordered vacancies. The samples were heated up to $T_c$ = 90 K.

The EDPs reveal two perpendicular sequences of reflections. The HREM images in the (100) and (010) sections could not be interpreted because of their complexity. A significant feature is the "incommensurability" in the EDP. In addition, in our experiments the grains of a polycrystalline aggregate of the 1-2-3 system with the same EDPs in (100) and (010) sections were observed. After exposure to an electron beam the precipitation of small particles of a new phase was observed. The ring-shaped EDP of this phase corresponds to the EDP of the original grain. We believe that the phase with the disordered heavy atoms begins to disintegrate under the influence of the electron beam; that is, that the small particles of the new phase are the particles of the usual ordered 1-2-3 phase. This interpretation confirms the explanation of the EDPs with the two mutually perpendicular sequences of long-periodic reflections in [2.6, 8]. It should be noted that the same transformation was observed on a background of twins, that is, in twinned grains with EDPs of the same type as above. In this case, before the formation of a ring-shaped EDP, that is, the formation of particles of a phase ordered with respect to barium and yttrium, the instability of the twin structure under the influence of the electron beam was observed before its complete disappearance. The instability was manifested by an alternate collapse of twins of one of the twinning directions with a corresponding change in the reflection spots on the EDPs. Since the majority of twinning models, discussed below in detail, contain an assumption about the "oxygen" nature of twin boundaries, the observed instability may be explained by the placement of barium and yttrium atoms along regular positions by the fluxes of oxygen vacancies, directed, most probably, along twin boundaries.

### 2.1.3 The Role of Oxygen in Formation of Twin Boundaries

*Barry* [2.9] studied the oxygen ordering by comparing high-resolution theoretical images to those obtained experimentally by means of an electron microscopy (JEM-400EX) in the multiple-wave approximation of relatively thick samples. The investigation demonstrated that it is possible to estimate the concentration of oxygen atoms on the condition that a relatively low-resolution experimental image is used along the bending direction of the (100) and (110) planes when they cross the twin planes (100). This bending was used to determine the ratio $a/b$ and to obtain indirect data for the degree of oxygen ordering in the O(4) sites. A significant discrepancy between individual measurements of the $b/a$ ratio and its average value of 1.016 was revealed. This may indicate that in samples of the 1-2-3 type, with complete ordering in Cu(1)-O(4) chains [where (1) and (4) are the positions of the atoms in the lattice, Fig.2.4] the $b/a$ relation exceeds the average value determined by neutron-diffraction experiments. It was also shown in the same work that after prolonged exposure to a 400 keV elec-

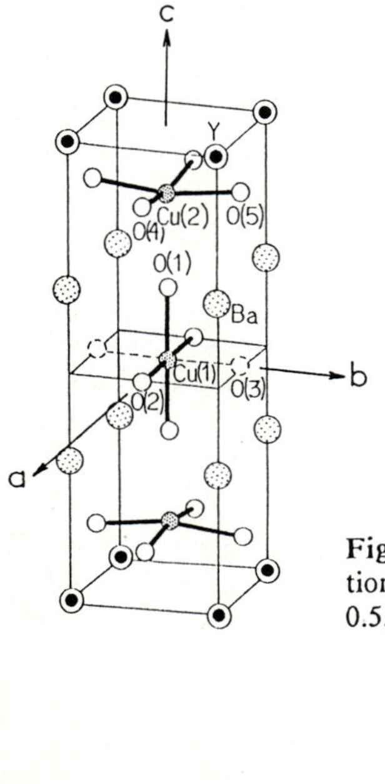


Fig.2.4. Crystal structure of $YBa_2Cu_3O_{7-x}$. The atom positions are Cu(1): 0,0,0; Cu(2):0,0,0.36; O(1): 0,0,0.16; O(2): 0.5,0,0; O(3): 0,0.5,0; O(4): 0.5,0,0.38; O(5): 0,0.5,0.38

tron beam, the HREM image of the twin planes vanishes. It is believed that the electron beam knocks oxygen atoms out of their equilibrium positions and that they are redistributed arbitrarily between the O(4) and O(5) sites, which leads to transformation of an orthorhombic structure into a tetragonal one.

A model of the emergence of coherent twin boundaries during a phase transition from the semiconducting tetragonal structure to the superconducting orthorhombic one has been proposed [2.10]. The model is based on electron-microscopy observations in situ [2.9, 11-13]. As was shown by neutron powder-diffraction measurements in situ [2.14] the O(4) and O(5) sites in the thermal region corresponding to the tetragonal structure are randomly occupied by equal amounts of oxygen atoms. The O(1), O(2) and O(3) sites are all fully occupied. It was demonstrated that the occupancy of the $(0,\frac{1}{2},0)$, *b*-chain sites takes place faster than the decrease of occupancy of the (1/2, 0, 0), O(5) sites. The similarity of the sites within each group and the retainment of the total amount of oxygen atoms means that the uptake of oxygen from the ambient atmosphere occurs primarily at the *b*-chain sites during oxygen ordering near the transition temperature. *Jou* and *Washburn* [2.10] suggested that the sudden increase of the rate of oxygen absorption is associated with the formation of nuclei of the orthorhombic phase. Since the oxygen-ordered orthorhombic phase contains linear chains on the basal plane, these nuclei are assumed to consist of clusters of short parallel *b*-chains, i.e., Cu(1)-O(4)-Cu(1). They should first appear as the heterogeneous sites at grain or pore surfaces, since a partial relaxation of internal stresses is likely to occur during the *b*-chains growth. The relaxation probably starts by consuming oxygen in the O(4)-O(5) basal plane, which thus becomes depleted near the grain boundaries. Because the initial tetragonal structure in the basal plane has the correct symmetry, the formation of the *b*-chain clusters is equally possible in the two mutually perpendicular directions. The same distribution of *b*-chain clusters, according to *Jou* and *Washburn*, fosters stress compensation induced by nucleation of an embryo. Thus, elongated embryos of twinning, surrounded by oxygen-depleted zones are expected. This model thus suggests a definition for the thickness of a twin boundary, namely, a plane of twin symmetry with an adjacent oxygen-depleted volume. The equilibrium width of the twin boundary is determined in this model by the balance of the repulsive forces of oxygen ions on opposite sides of the boundary and the chemical potential favoring extension of the *b*-chains.

The contrast in HREM images was simulated under various defocusing conditions and different foil thicknesses as a function of the number of oxygen-depleted layers neighboring the plane of a twin symmetry. It was demonstrated that, when there are several such layers, contrast occurs across the boundary. Various types of defect images have indeed been observed across coherent twin boundaries [2.7, 15-17].

However, the strong influence of defocusing upon such contrasts, as well as alternative interpretations of the data must be considered. For example, a model of twin boundaries as transitional diffusive zones along

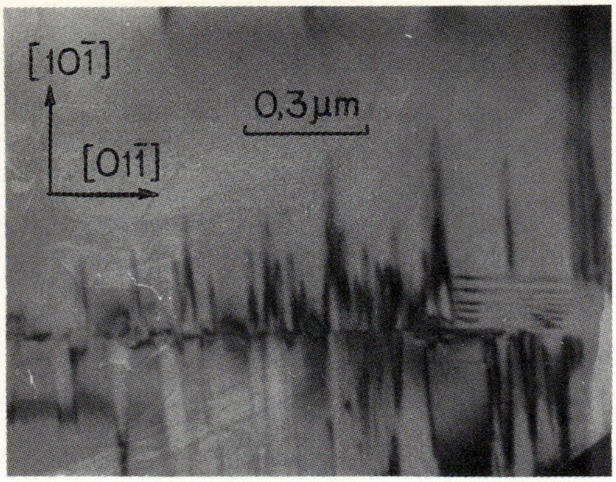


**Fig.2.5.** Twins and nuclei of twinning in a crystal of $YBa_2Cu_3O_{7-x}$

which the parameters of one twin orientation transform into those of the other continuously and smoothly is a possibility discussed in Chap.3. There, coherent and incoherent boundaries are explained: A coherent boundary also has a finite thickness, estimated to be less than 100 Å, and may yield a corresponding contrast in the HREM image. Furthermore, this interesting model (named the "oxygen" mechanism) also requires continuous uptake of oxygen from the ambient atmosphere to sites of embryo formation and also rapid diffusion of oxygen atoms in the bulk. These processes might easily occur in polycrystalline ceramic materials because of the large number of pores and intergranular boundaries along which the diffusion of oxygen is restricted.

According, however, to our electron-microscopy investigations, the distribution of twin layers in poly- and single crystals is more-or-less homogeneous, although the conditions for consuming oxygen from the ambient atmosphere and for its rapid diffusion in single crystals, particularly in large ones, are stricter. Moreover, investigation of the junction of the two twin systems, which according to the model can be interpreted as the zone of formation of twin layers in which lentil-shaped nuclei can be seen (Fig. 2.5), revealed that such nucleation regions easily occur deep inside a single-crystal volume.

We believe that the formation and spread of twin dislocations are the basic factor determining the kind and configuration of twin layers, although the importance of the oxygen-ordering processes with consequent generation of *b*-chain conglomerates for the formation of twin layers should not be neglected. In electron microscopes, the twin layers are normally observed at a temperature lower than that required for ordering of oxygen atoms. Figure 2.5 shows twin layers neighboring the small angle boundary in a 1-2-3 single crystal. Along with lentil twin layers formed probably via the "oxygen" mechanism discussed above, in the vicinity of the small angle boundary which serves as a channel for oxygen diffusion, twin ejections can be seen projecting directly from the small-angle boundary. By cancelling the contrast of the various reflections, the presence of dislocations in the walls of twin wedges could be revealed. These coincide with the

Burgers vector in the $[1\bar{1}0]$ direction. The similar size of the two types of twin layers indicate that the formation and spread of twin nuclei associated with the oxygen mechanism takes approximately the same time as the formation of wedge-shaped twins by ejection of twinning dislocations. Hence, these two mechanisms must be directly competing and adding to each other already at the starting stages of twin-structure formation.

Dislocations with the $\langle 110 \rangle$ Burgers vector, which appear to be twinning, were discovered after deformation of an orthorhombic phase at room temperature [2.18, 19]. *Van Tendeloo* et al. [2.20] used twinning dislocations with a $\tau = b\text{-}a$ shear to describe the shape of free edges of twin layers in terms of a dislocation model. These dislocations are also required to explain the experiments on detwinning of crystals, in which non-axial, compressing mechanical stress is applied along the [100] or [010] directions of a perovskite structure at 400° to 450° C at a partial pressure of oxygen sufficient to support stability of the 1-2-3 phase [2.21]. Interestingly, although there are plenty of states and junctions of twin systems available under ordinary growth conditions for single and polycrystals of $YBa_2Cu_3O_{7-\delta}$, a so-called *tweed texture*, that is, an equally probable overlapping of narrow microtwins, is always observed after exposure of a foil to an electron beam or after heating in situ. The microtwins do not cause splitting of long-range reflections on the EDP and show the average symmetry of the fourth-order EDP by cross-shaped diffuse-reflection spots. We note that the formation of a tweed texture in situ occurs when there is an oxygen deficiency, i.e., for a thin foil, in vacuum, or under illumination by an electron beam. Thus, the tweed structure is the ideal realization of homogeneous ordering of oxygen atoms under oxygen-poor conditions.

The HREM study of the tweed structure was described in [2.22]. The tweed image consists of numerous mutually perpendicular overlapping domains of small extent with somewhat diffuse boundaries, positioned approximately parallel to the traces of the $(1\bar{1}0)$ and $(110)$ planes.

The lattice displacement **R** appears to be of a shear type on the $\{110\}$ planes in the $\langle \bar{1}10 \rangle$ directions, as indicated by experiments measuring the contrast. HREM showed that the modulation of atomic levels has a period of 3÷5 nm. The analysis of the modulation, together with the diffusion of reciprocal lattice junctions, led *Zhu* et al. to conclude that long-range static deformations of the $\{110\}$ $\langle \bar{1}10 \rangle$ type exist. It is interesting to note that the tweed structure was observed both in samples with an acute oxygen deficiency $YBa_2Cu_3O_{7-\delta}$ ($\delta \geq 0.6$) and in substituted samples $YBa_2(Cu_{1-y}M_y) \cdot O_{7+\delta}$ (M = Fe, Co, Al, Ga; $y < 0.02$).

## 2.2 The Structure of Defects in the Bi-Sr-Ca-Cu-O System

### 2.2.1 Structural Modulation

In 1988, after the discovery of high-temperature superconducting Bi-Sr-Ca-Cu-O compounds with $T_c \simeq 110$ K, the so-called high-temperature phase $Bi_2Sr_2Ca_2Cu_3O_x$ (named 2-2-2-3) ($x \simeq 10$, $T_c = 110K$) and low-temperature phase $Bi_2Sr_2CaCu_2O_x$ (named 2-2-1-2) ($x \simeq 8$, $T_c = 80K$) with higher critical current were distinguished. The average crystal structure of a $Bi_2Sr_2CaCu_2O_x$- type compound was put forward by *Tarascon* et al. [2.23] as a tetragonal cell with $a = b = 0.54$ nm and $c = 3.08$ nm. The structure consists of alternate stacking of $Bi_2O_2$ layers and perovskite layers along the $c$-axis. Each $Bi_2O_2$ layer is composed of two BiO planes with bismuth and oxygen atoms distributed as in an NaCl crystal. Every perovskite layer consists of two $CuO_2$, two SrO, and one Ca layers. The general sequence along the $c$-axis is

BiO | -SrO-$CuO_2$-Ca-$CuO_2$SrO- | | BiO-BiO- | SrO

↓ perovskite ↓ $Bi_2O_2$

A higher-temperature $Bi_2Sr_2Ca_2Cu_3O_x$ phase in perovskite layers contains additional layers of Ca and $CuO_2$ with the parameter of the $c$-lattice increased to 3.6 nm. The lowest-temperature phase, $Bi_2Sr_2CuO_x$ ($T_c = 30K$), is obtained by removing the sequence of calcium and $CuO_2$ layers with $c$ decreased to 2.4 nm.

HREM revealed strong deviations from the so-called *basis structure* induced by a modulated structure which is often incommensurate with the basis structure, in all compounds of the bismuth system. The modulated structures are difficult to study by conventional X-ray and neutron diffraction techniques, so most of the information was obtained by electron-microscopic methods. Modulated structures have been seen in all phases of the Bi-Sr-Ca-Cu-O systems. In particular, the HREM of the $Bi_2Sr_2Ca_2 \cdot Cu_3O_x$ ($T_c = 110K$) high-temperature phase obtained by an incident beam along the [110] direction reveals perovskite layers of various thickness with an "average" tetragonal strfucture with $a = 0.54$ nm, $c = 3.6$ nm [2.24]. Of the observed bands of perovskite layers, 70% consist of three Cu layers, corresponding to the average-structure model. However, really 20% of them contain three Cu layers and 10% have four Cu layers. Within the microscope's field of view the inhomogeneous occupation of the perovskite layers in this section appeared to be retained over the entire "perovskite" band.

In the [100] section, besides the various thicknesses of the perovskite layers along the $c$-axis, each of them containing 2, 3, 4 or 5 "copper" sublayers, alternating regions of higher and lower density of $Bi_2O_2$ layers in the $b$-direction were revealed. In this section the contrast of bismuth layers on the HREM images was higher than for other layers. This permitted us to estimate the period of the inhomogeneity in the concentration of bismuth atoms as 2.7 nm. The maximum deviation of the transverse length of the perovskite layer from the average length as seen as on the HREM lattice image is ~20%.

The (001) section of the same material was also studied [2.25]. When the electron beam is directed along $c$, the contrast of the HREM image is low and only faint dark bands extending along $a$, perhaps associated with layers of higher-density bismuth atoms, could be seen. No bands of higher concentration are observed in the (510) plane, although according to the geometry they should be seen [2.24]. This has not been understood so far. All these observations suggest that the $Bi_2O_2$ layers do not represent two perfect two-dimensional layers, but rather contain fluctuations of concentration and position of Bi atoms within the layers. This may be due to either the substitution of Bi sublattice sites with atoms of Ca and Sr, or to some other mechanism which induces distortion of the lattice by shifting the positions of the Bi atoms, without substitution. In any case, the $Bi_2O_2$ layer is split into bands which are either deficient in or enriched with Bi atoms, alternating along $b$. Each concentration band of Bi is extended along $a$ with no marked variation in width, and along $b$ it is approximately 1.5 nm thick. The layers are arranged so that the denser regions of Bi atoms are positioned above the less-dense volumes.

According to *Matsui* et al. [2.25], this leads to the formation of a body-centered lattice. Although the period of the regions of higher density of Bi atoms in a [100] section is usually 10 lattice planes, or 2.7 nm in the $b$-direction, sometimes a period of 9 lattice planes (2.4nm) is observed. Such a shear along the $b$-axis may be the cause for the observed incommensurate structure where $b$ is approximately $4.8a$ (2.6nm).

A computer simulation of the HREM image which assumed that the $B$-site of perovskite layers are substituted by Cu atoms and the $A$-sites by Sr or Ca atoms, showed qualitative agreement with the observed images [2.25]. The crystallography of the modulation structure in the lowest-temperature phase of $Bi_2Sr_2CuO_x$ (T = 20K) strongly differs from that existing in other phases [2.26]. Analysis of the HREM images and EDPs in the [100] section demonstrates that the nuclei of Bi concentration fluctuation form a monoclinic superlattice with $\mathbf{A}$ = 0.54 nm, $\mathbf{B}$ = 2.6 nm, $\mathbf{C}$ = 2.8 nm and $\alpha = 116°$, but that it is not body-centered as in the previous case. This means that the centers of higher Bi-atom density both in the upper and lower BiO planes are slightly shifted along the $b$-direction; this was not observed in either $Bi_2Sr_2Ca_2Cu_3O_x$ or $Bi_2Sr_2CaCu_2O_x$. The value of the shift is estimated as $b/4$.

The modulated structure $Bi_2Sr_2CaCu_2O_x$ has been studied by HREM methods [2.27]. A regular tetragonal sublattice was seen in the cross section when modulations were absent, according to *Tarascon* [2.23]. Additional reflections associated with modulation were also absent in the EDPs. However, a distinct modulation in the concentration of Bi atoms in a $Bi_2O_2$ layer in the $b$-direction was observed in the (100) section and additional reflections were found in the corresponding EDP. The superlattice of centers of increased Bi concentration is body-centered orthorhombic with $A = a =$ 0.54 nm, $B = 4.8b =$ 2.6 nm and $C = c =$ 3.08 nm. In this section it is clearly seen that the vertical planes of the lattice occupying the space between

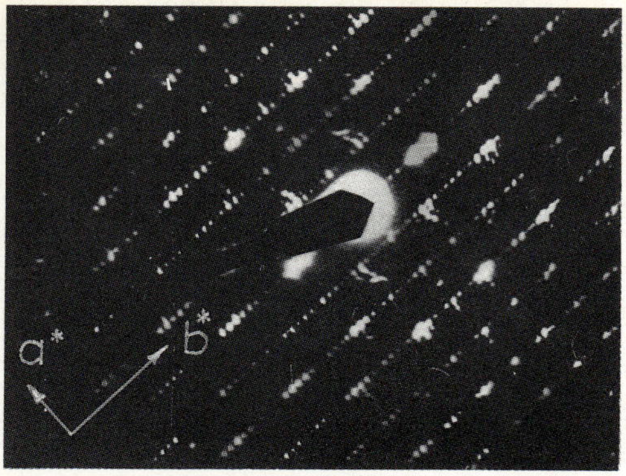


Fig.2.6. Electron diffraction pattern of modulated $Bi_2Sr_2CaCu_2O_{8+x}$. $a^*$ and $b^*$ are the axes of the reciprocal lattice. The bright spots are reflections corresponding to junctions of the reciprocal lattice. The lattice modulation results in the appearance of weaker, additional reflections (as compared with the main reflections) along the $a^*$ and $b^*$ axes

$Bi_2O_2$ layers and expected to be straight, are periodically zigzag; the distance between them is 0.27 nm.

The simultaneous existence of the modulation with twin planes (110) and the twin boundaries (001) was described in [2.28]. The twin-symmetry planes are positioned in the center of $Bi_2O_2$ layers. Our HREM measurements carried out on crystals of this composition [2.29] showed low-contrast modulation bands in the cross section of the (001) plane. The HREM images show fragments of lines perpendicular to the main modulation direction. The EDPs obtained from single crystals of this composition (Fig.2.6) can often exhibit short, broken rows of reflections (complementary to matrix reflections) perpendicular to additional modulation reflection rows.

In order to avoid the error associated with a possible superposition of two monocrystalline blocks, conventional electron microscopic procedures have been carried out. In addition, it should be noted that the rows of perpendicular reflections were observed only for modulation reflections and never for the basic matrix reflections. This may imply that the regions bounded by maxima of waves of high and low density in the $Bi_2O_2$ layers observable in the (100) plane are not always linear in the direction perpendicular to this plane. If the period of the main modulation in the (100) section is 2.7 nm, the modulation in the perpendicular direction has a smaller period of about 1.2 nm, which accounts for an insignificant fraction of the main modulation. Interestingly, both modulations exist simultaneously without any noticeable boundary between them. In contrast to *Zhu* et al. [2.30], we cannot regard these data as an indication of the existence of a two-dimensional modulation in this plane, since fragmentary microtwinning is also possible. On the other hand, a geometrical model of the two-dimension modulation must be three-dimensional, i.e., it has to comply with observations in two other mutually perpendicular cross sections. It is possible that this is simply due to errors in the packing of modulation layers similar to those observed in the (010) section [2.31].

### 2.2.2 Models of Structural Modulation

All the data collected by electron microscopy indicate that the main cause of structural modulation is due to specific features of the structure of the double $Bi_2O_2$ layer, which give rise to waves of alternating regions of high and low density. *Zandbergen* et al. [2.31] have analyzed the possible causes of these peculiarities, the choice among which, judging from the published data, is difficult to make:

(i) **Excess oxygen in the $Bi_2O_2$ layers.** Over-saturation with oxygen of BiO layers forces the unit-cell parameters $a$ and $b$, which are rigidly set by the perovskite layer structure, to be too large (from the viewpoint of configurational chemistry) for the Bi atoms to fall into a linear arrangement along the BiO layers. Accommodation of this layer may lead to the formation of vacant sites for the oxygen atoms. Data giving a mean valence of metal atoms in a compound exceeding the normal one support this hypothesis [2.23, 32]. The force which drives the formation of density waves in this case is not the establishment of charge balance with the extra oxygen but a better conjugation of two $Bi_2O_2$ layers and a perovskite layer. The main disadvantage of this model is the change required in the Bi-Bi spacing which has so far not been seen on high-resolution images.

(ii) **Sr atom deficiency.** It is believed that for $Bi_2Sr_2CaCu_2O_8$ the modulation is caused by a periodic absence of Sr atoms and a corresponding atomic redistribution [2.33]. This causes the $CuO_5$ to tilt, thus leading to a modulation.

(iii) **Atomic substitution.** Partial substitution of Bi by Cu (one atom out of ten) (also one out of ten atoms) gives rise to structural modulation.

(iv) **Changes in the orientation of bismuth lone pairs.** $Bi^{3+}$ cations have a lone pair of electrons that often leads to an asymmetry in the coordination sphere in bismuth compounds. Thus, modulation in BiO layers may be attributed to variations in the orientation of the lone pairs. Such variations can arise either spontaneously or upon implantation of extra oxygen atoms into the BiO layer.

(v) **Combination of iii and iv.** A simultaneous implantation of extra oxygen into the $Bi_2O_2$ layer and a partial substitution of Bi by Cu and Sr by Bi may also give rise to modulated structures.

Another model for the modulation based on the ordering of oxygen vacancies [2.34] was rejected by *Zandbergen* et al. [2.31] on the grounds that no changes in the modulated structure were observed up to the melting point and, therefore, there are no phase transitions in this region.

In the same paper, model (iv) regarding the bismuth lone pairs has also been criticized. The structure with ordered lone pairs is known in the literature as the *Aurivillius structure*, which in nonsuperconducting bismuth compounds exhibits a phase transition at 200° -400° C to a state where the lone pairs are disordered. The *Zandbergen* et al. believed that the order-disorder transition is a necessary feature of the Aurivillius state of a similar layer that is formed upon conjugation of two bismuth layers [2.23,35-36]. A model of the modulation structure based on an Aurivillius state of a similar oxide layer has also been described in [2.37].

It should be noted that the criticism of these models was based on electron-microscopic data obtained from powder samples deposited onto a substrate, usually carbon. A significant point of the criticism is the absence of changes in the modulation upon heating and therefore the absence of phase transitions. Specific features of changes in the structure under in situ electron irradiation were observed in [2.38]. However, *Zhang* et al. did not attribute these changes to a phase transition. Our work was performed on $Bi_2Sr_2CaCu_2O_{8+x}$ single crystals [2.29]. HREM samples were prepared by cutting thin pellets with subsequent thinning by chemical means. It was found that 100 keV electron bombardment does not lead to structural changes. However, JEM-4000EX-microscope studies revealed that 400 keV electron bombardment does give rise to new structural states, which are described below:

(i) The matrix symmetry in the section (001) plane is retained and some modulation on HREM images disappears. Spots of "grey" contrast show up, modulation reflections disappear and the character of the EDP changes (Fig.2.7).

(ii) As bombardment continues, a new phase with a different lattice symmetry and larger interplanar spacing forms within the grey contrast spots.

Structural changes upon in-situ heating in a JEM-100CX microscope were studied in [2.39]. Combination of fast and slow heatings revealed two temperature regions in which structural changes, observed in the EDP, take place. These changes are interpreted as second-order phase transitions. One

Fig.2.7. HREM image of the $Bi_2Sr_2CaCu_2O_{8+x}$ single crystal in the (001) section. Nucleation of a new phase under an electron beam can be seen at the left

of these temperature regions is precisely between 200° C and 400° C. It is interesting that modulation in the local region of observation also disappears, i.e., in order for it to disappear, simultaneous electron bombardment and heating is required. Many experiments demonstrated that heating, together with irradiation in vacuum, leads to slow loss of oxygen from samples. Therefore, perhaps the model based on the change of orientation of bismuth lone pairs under the effect of excess oxygen in BiO layers should be taken as the basis for the actual structural modulation model.

Other defects characteristic for the basal plane of Bi-Sr-Ca-Cu-O crystals are dislocations and dislocation networks. Work carried out with such crystals doped with Pb reveals that after addition of Pb the structure of a 2-2-2-3 unit cell, in which 30% of Bi is replaced by Pb, appears [2.40, 41]. The main gliding plane is considered to be the basal one. In experiments measuring contrast, *Li* et al. [2.40] and *Takahashi* et al. [2.41] discovered dislocations with Burgers vectors in the directions [100] and [010], forming hexagonal networks. It was assumed that dislocations with the [110] Burgers vector are not stable and are divided into two partial ones. Splittling of dislocations with [100] Burgers vectors was also observed. Experiments measuring contrast, conducted in the basal plane of 2212 single crystals, showed that the observed dislocation networks consist of long branches with the Burgers vectors directed along [100] and [010] and divided into partial dislocations with rather broad stripes of a stacking fault (to ~800 Å). Interaction of these dislocations results in a new stable dislocation with the Burgers vector directed along [110] and with the stacking-fault width several times less than that for dislocations with Burgers vector along [100] and [010] (Fig.2.8). Thus all the observed stacking faults lie in the (001) basal plane. *Eible* [2.42] surveyed the stacking faults in 2212 crystals by HREM. These faults, not being associated with any definite crystallographic attitude plane,

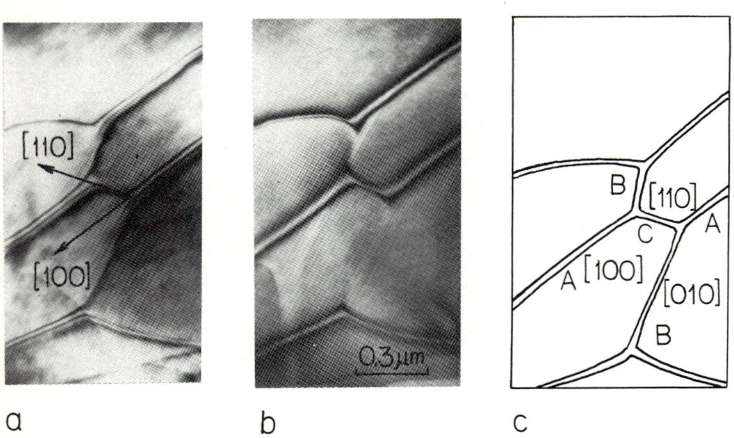


**Fig.2.8.** Contrast experiments carried out on dislocations in an as-grown specimen. The direction of the incident beam is approximately parallel to the c-axis. (**a**) Fragment of the dislocation network in **G** = [220]. (**b**) One of the experiments in which in **G** = [200] one branch of dislocations denoted by B disappears. (**c**) Schematic illustration of a fragment of the dislocation network

are no more than ~2 nm wide and separate two regions of different chemical composition. Because they are extended faults in the chemical composition, according to *Eible*, they facilitate solid-state diffusion along the *c* axis. Moreover, the stacking faults not associated with any definite crystallographic plane fix twin boundaries, which explains their frequent presence in the region of twin boundaries. Sometimes stacking faults form complex conglomerates with twin boundaries, the plane of the stacking fault may be placed even over a cylindical surface irrespective of the crystallographic plane being investigated. This proves that the plane of the stacking faults can easily change its internal structure and that the defect energy is insignificantly influenced by the orientation of the defect attitude plane. Here we mean the stacking faults of the "chemical type" described by *Eibl*. As to the ordinary stacking faults lying in the basal plane, they are characterized by the interesting fact that under appropriate diffraction conditions, the modulation stripes in the basal plane, together with the stretched sites of dislocation reactions, may be observed simultaneously. "Modulation" planes, positioned perpendicularly to the attitude plane of a stacking fault, pierce the latter and the partial dislocations without noticeable interaction or accommodation effects.

We have measured in situ the width of a stacking-fault stripe with the Burgers vector along [100] as a function of temperature. It was noted that the width of a stacking fault decreases monotonically between 100° C and 350° C as oxygen is being removed. This occurs in a thin sample heated in vacuum. No theoretical simulation of this behavior of stacking faults of either the ordinary type, lying in the basal plane, or of those causing "chemical shear" has been done to date. However, the experimentally obtained data permit us to draw a conclusion not only about the oxygen nature of the modulation structure, but also about the leading role played by oxygen in the formation of a stacking fault stripe in a basal plane. Indeed, the visible decrease in the width of faults with increasing temperature while the modulation structure is retained, and the absence of accommodation effects when modulation planes intersect fault planes, can be explained only in a similar way.

Although it has been established that oxygen atoms are important in determining the properties of structural defects in superconducting materials, there is still a great deal to be done to devise working "oxygen" models for defect structure and to understand its influence on superconductivity parameters.

## References


2.1 Y. Matsui, E. Takayama-Muromachi, K. Kato: J. Appl. Phys. 27, L350-L353 (1988)

2.2 T. Siegrist, S. Sunshine, O.W. Murphy, R.J. Cava, S.M. Zahurak: Phys. Rev. B **35**, 7137 (1987)

2.3 M.A. Beno, L. Soderholm, D.W. Capone II, D.G. Hinks, J.D. Grace, I.K. Schuller: Appl. Phys. Lett. **51**, 51 (1987)

2.4 Y. Matsui, E. Takayama-Muromachi, A. Ono, S. Horiuchi, K. Kato: Jpn. J. Appl. Phys. **26**, L777 (1987)

2.5 S. Ikeda, T. Hatano, A. Matsushita, T. Matsumoto, K. Ogawa: Jpn. J. Appl. Phys. **26**, L729 (1987)

2.6 H.W. Zandbergen, G. van Tendeloo, T. Okabe, S. Amelinckx: Phys. Stat. Sol. (a) **108**, 45-71 (1987)

2.7 G. van Tendeloo, H.W. Zandbergen, S. Amelinckx: Solid State Commun. **63**, 603-606 (1987)

2.8 G. van Tendeloo, H.W. Zandbergen, T. Okabe, S. Amelinckx: Solid State Commun. **63**, 969-972 (1987)

2.9 J.C. Barry: J. Electron. Micros. Tech. **8**, 325-337 (1988)

2.10 C.J. Jou, J. Washburn: J. Mater. Res. **4**, 795-801 (1989)

2.11 G. Van Tendeloo, S. Amelinckx: J. Electron Microsc. Techn. **8**, 285 (1988)

2.12 M. Sugiyama, R. Suyama, T. Inuzoka, H. Kubo: Jpn. J. Appl. Phys. **26**, L1202 (1987)

2.13 T.E. Mitchell, T. Roy, R.B. Schwarz, J.F. Smith, D. Wohlleben: J. Electron. Microsc. Tech. **8**, 317 (1988)

2.14 J.D. Jorgensen, M.A. Geno, D.G. Hinks, L. Soderholm, K.J. Volen, R.L. Hittenman, J.D. Grace, I.K. Schuller, C.U. Segre, K. Zhang, M.S. Kleefisch: Phys. Rev. B **36**, 3608 (1987)

2.15 K. Hiraga, D. Shindo, M. Hirabayashi, M. Kikuchi, Y. Syono: J. Electron. Microsc. **36**, 261 (1987)

2.16 G. Van Tendeloo, H.W. Zandbergen, S. Amelinckx: Solid State Commun. **63**, 389 (1987)

2.17 Y. Hirotsu, Y. Nakamura, Y. Murata, S. Nagakura, T. Nishihara, M. Takata: Jpn. J. Appl. Phys. **26**, L1168 (1987)

2.18 J. Rabier, M.F. Denanot: J. Less-Common Metals **164**, 223-230 (1990)

2.19 V.S. Bobrov, V.A. Goncharov, G.A. Emel'chenko, A.P. Ivanov, R.K. Nikolaev, A.N. Izotov, J.N. Novomlinskii, Yu. A. Ossipyan, N.S. Sidorov, E.V. Suvorov, V.Sh. Shekhtman, L.N. Zavel'skaya, I.I. Zver'kova: Proc. 7th Sintec World Ceramic Congress, Sattelite Symp. HTSC, Trieste **8**, 49-54 (1990)

2.20 G. Van Tendeloo, D. Broddin, H.W. Zandbergen, S. Amelinckx: Physica C **167**, 627-639 (1990)

2.21 V. Welp, M. Grimsditch, H. You, W.H. Kwok, M.M. Fang, G.W. Grabtree, J.Z. Liu: Physica C **167**, 1 (1989)

2.22 Y. Zhu, M. Suenaga, A.R. Moodenbaugh: Phil. Mag. Let. **62**, 51-59 (1990)

2.23 J.M. Tarascon, Y. Le Page, P. Bardoux, B.G. Bagley, L.H. Green, W.R. McKinnon, G.W. Hull, M. Giroud, D.M. Huang: Phys. Rev. B **37**, 9382 (1988)

2.24 Y. Matsui, S. Takekawa, H. Nozaki, A. Umesono, E. Takayama-Muromachi, S. Horiuchi: Jpn. Appl. Phys. **27**, L1241-L1244 (1988)

2.25 Y. Matsui, H. Maeda, Y. Tanaka, Sh. Horiachi: Jpn. J. Appl. Phys. **27**, L372-L375 (1988)

2.26 Y. Matsui, H. Maeda, Y. Tanaka, Sh. Horiuchi, Sh. Takekawa, E. Takayama-Muromachi, A. Umezono, K. Ibe: JEOL. News **26**, E-21 (1988)

2.27 Y. Matsui, H. Maeda, Y. Tanaka, S. Horiuchi: Jpn. J. Appl. Phys. **27**, L372-L375 (1988)

2.28 Y. Matsui, H. Maeda, Y. Tanaka, E. Takayama-Muromachi, S. Takekawa, S. Horiuchi: Jpn. J. Appl. Phys. **27**, L827-L829 (1988)

2.29 V.A. Gontcharov, A.B. Ermolaev, L.N. Zavel'skaya, Yu.A. Ossipyan, E.V. Suvorov: Metallophysics **13**, No.4, 32-39 (1991)

2.30 J. Zhu, H.Q. Ye, X.W.Lin, A.Q. He, S.Q. Feng, X. Zhu, Z.Z. Gan: Mod. Phys. Lett. B **3**, 145-150 (1989)

2.31 H.W. Zandbergen, W.A. Groen, F.C. Mijlhoff: Physica C **156**, 325-354 (1988)

2.32 T. Kajitani, K. Kusaba, M. Kikuchi, N. Kobayashi, Y. Syono, T.B. Williams, M. Hirabayashi: Jpn. J. Appl. Phys. **27**, L587 (1988)
2.33 P.L. Gai, P. Day: Physica C **152**, 335 (1988)
2.34 Y. Syono, K. Hiraga, N. Kobayachi, M. Kikuchi, K. Kusaba, T. Kajitani, D. Shindo, S. Hosoya, A. Tokiwa, S. Terada, Y. Moto: Jpn. J. Appl. Phys. **24**, L569 (1988)
2.35 M.A. Subramanian, C.C. Torardi, J.C. Calabrese, J. Gopalakrishnan, K.J. Morrisey, T.R. Askew, R.B. Elippen, U. Chowdhry, A.W. Sleight: Science **239**, 1015 (1988)
2.36 H.G. von Schnering, L. Waltz, M. Schwartz, W. Becker, M. Hartweg, T. Popp, B. Hettich, P. Muller, G. Kampf: Angewandte Chem. **27**, 547 (1988)
2.37 X. Zhu, O. Yin, B. Lih: Mod. Phys. Lett. B **4**, 59-62 (1990)
2.38 X.F. Zhang, Y.F. Yan, K.K. Fung: Phil. Mag. Lett. **60**, 11-16 (1989)
2.39 V.A. Gontcharov, E.Yu. Ignat'eva, Yu.A. Ossipyan, E.V. Suvorov: Metallophys. (1992) (in press)
2.40 Z.-Q. Li, H. Shen, Y. Qin, J.-Y. Jiang, J.-J. Do: Phil. Mag. Lett. **60**, 123-130 (1989)
2.41 Y. Takahashi, M. Mori, Y. Ishida: In Proc. ISNC'89, Tokyo, Jpn., pp.330-333
2.42 O. Eibl: Physica C **168**, 249-256 (1990)

# 3. Twins, and Structure of Twin Boundaries

I.M. Shmyt'ko and V.Sh. Shekhtman

The twinning parameters for crystals of the $RBa_2Cu_3O_7$ system, R being a rare-earth element (Y, Gd, Eu, Ho), and the $La_2CuO_4$ system were determined by X-ray methods. It appeared that twin boundaries in a 1-2-3-$O_7$ system are transitional zones inside which the parameters of one twin orientation pass over to the characteristics of the other twin orientation continuously and smoothly. The thickness of the transitional zones for coherent and incoherent boundaries varies significantly. It can be up to several twin thicknesses for incoherent conjugations and less than 100 Å for coherent ones. This structure is determined not by a disturbance in the stochiometric composition but by a continuous change in the degree of ordering in the oxygen atoms along the *a* and *b* directions, their concentration being virtually constant. Rapid cooling of 1-2-3-$O_x$ crystals leads to the formation of a quasi-twin structure in which no distinguishable twin orientations are observed. However, a quasiperiodic change of rhombicity angle occurs so that the crystal consists of only transitional zones. Such a state is also determined by a change in the degree of ordering of the oxygen atoms along the x and y axes, their concentration being constant.

## 3.1 History of the Problem

Twins are a type of substructure which may arise during crystal growth, recrystallization, exposure of a crystal to mechanical forces and during phase transitions, when so-called *transformation twins* occur. In the latter case the twins are formed in different orientation states, the number of which is defined by the change in the lattice symmetry during the phase transition. The distribution of the orientation states is determined by conditions of the transformation, the history of the sample, structural defects and so on. Thus, the complicated substructure of a polytwin crystal cannot be interpreted as a completely asymmetric configuration. The Curie principle of conservation of macrosymmetry should be taken into account [3.1-5]. Accordingly, the first experimental question to be investigated is the number and configuration of the orientation states and their complexes.

The second problem is related to the regions between adjacent twins. Coherent and incoherent intertwin boundaries have been observed. Figure 3.1 shows the configurations giving these two types of boundaries and the twin complexes. Each such complex may contain a great number of twin domains with a common type of twin plane.

Springer Series in Materials Science, Vol. 23
**The Real Structure of High-$T_c$ Superconductors**
Editor: V.Sh. Shekhtman 

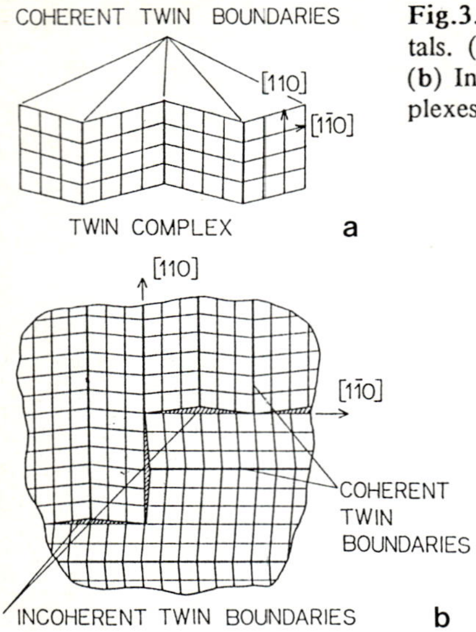


**Fig.3.1.** Twin configurations in 1-2-3-$O_{7-\delta}$ crystals. (**a**) Twin complex with coherent boundaries. (**b**) Incoherent twin boundaries between twin complexes with perpendicular twin planes

The simple assumption of a twin boundary describes it as an ideal monatomic plane passing through the symmetry plane of a highly symmetric phase which is skipped at the boundary between two phases. However, consideration of a polycomponent system with twins reveals that, using this assumption, abnormal situations occur in which the conjugation of twins in the same region of space requires the presence of several sorts of atoms. Examples of such superpositions of atoms are presented graphically in the book on mechanical twinning by *Klassen-Neklyudova* [3.6]. This contradiction can be eliminated by assuming that the regions of twin conjunction are extended. The theoretical study of twin boundaries as transitional zones was first performed by *Kontorova* in 1942 [3.7]. She proceeded from the assumption that the interaction energy between layers of different orientation and the energy of "orientational forces", determined by the departure of atoms from ideal positions in terminal twin orientations, balance one another. The sum of these energies will depend on the transition zone width, which is on the order of a few hundred interatomic spacings, as obtained by *Kontorova*.

Historically, the existence of extended intertwin boundaries was accepted only for ferroelectrics [3.8-11]. The existence of transitional zones in ferroelastics was not studied experimentally. Such zones must exist in high-temperature superconductors. Critical currents in the 1-2-3-$O_x$ system have significant angular dependence even when passing through an intergrain boundary [3.12-15]. Thus, wide transition zones in the region of twin boundaries must lead to an analogous effect. Such a supposition is based on an experimental study of the pinning properties of twin boundaries for superconducting vortexes [3.16-18]. For this reason, twin structure and the structure of twin boundaries in HTSC crystals should show the relation between the actual structure and the physical properties of these compounds.

## 3.2 Experimental Technique

The primary method used in this work was angular scanning X-ray topography and precision diffractometry. The first method was successful in studying the twinning in single crystals of $BaTiO_3$ [3.19] and $KH_2PO_4$ [3.10]. A standard X-ray diffractometer is sufficient. In our case an X-ray installation of the DRON make was employed.

Two methods of scanning are possible: $\Theta$-$2\Theta$ and $\omega$. In the first, the sample is rotated with the angular velocity $\omega$ along with a photographic film (placed perpendicularly to the axis of the diffracted beam, in front of the detector input slit) rotating at twice the velocity $2\omega$. In the second variant only the sample rotates.

In the first case the method is sensitive to a misorientation of the mosaic blocks around the axis parallel to that of the goniometer rotation, but it is not sensitive to changes in lattice parameters. In $\omega$-scanning the misorientation of blocks in the direction of scanning is not registered on the topogram, but the reflections corresponding to various interplanar distances are fixed. In both schemes, misorientations of sample fragments around the axis perpendicular to that of the goniometer rotation are related to the picture obtained along the vertical axis.

It is presumed that the diffracting region on a sample is very small and does not distort the topographic image. To fulfill this condition, narrow collimating slits are required. In $\omega$-scanning it should be taken into account that the image is superimposed on the continuous spectrum of X-ray beams. The essential requirement on the X-ray source is that the size of the emitting region be minimum (in our case the source size was $50 \times 5.0\,\mu m$).

## 3.3 $YBa_2Cu_3O_{7-\delta}$ and $La_2CuO_4$ Crystals

The twinning parameters for 1-2-3 crystals were determined from the splitting of reflections on Laue patterns and oscillating X-ray patterns. Figure 3.2 depicts Laue patterns on which it is seen that the planes where the zone axis is parallel to [001] yield the splitting of reflections along the perimeter of the zone ellipse. In contrast, the reflections belonging to the [100] zone (Fig.3.2b) are split across the perimeter of the zone ellipse. It is seen that reflections of the type (001) for all orders remain focused unsplit dots. All these feature of diffraction are obtained directly during detailed crystallographic analysis using an assumed lattice transformation following the scheme of a homogeneous shear. The basic conclusion is that a new structure is formed by twinning of the initial tetragonal lattice according to the system $\{110\}/\langle 100\rangle$.

The scheme shown in Fig.3.3 illustrates the transition to a rectangular cell by tilting the parent square cell OABC by an arbitrary angle. In this variant of a twin shear a new phase is formed in the Orientation State (OS) (Fig.3.3b). The mirror-symmetric domain in the figure is denoted as II.

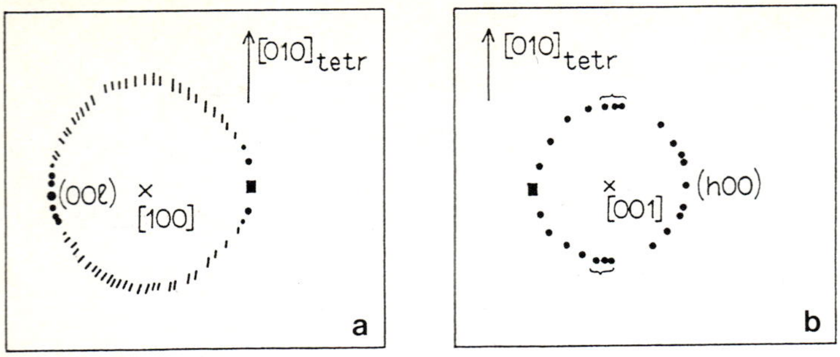


**Fig.3.2.** Schematic Laue patterns of 1-2-3-$O_{7-\delta}$ single crystals: (**a**) the axis of the reflection zone runs parallel to $[100]_{tetr}$ of the tetragonal phase; (**b**) the axis of the reflection zone is parallel to $[001]_{tetr}$. The dots and dashes give the shapes of the Laue reflections. The square denotes the primary beam

Alternatively, the twinning domains III and IV can occur. Figures 3.3c,d depict the reciprocal lattices of the initial tetragonal phase and the resulting orthorhombic one which is a superposition of the four orientation states I-IV. Note that in this scheme with homogeneous deformations, a transition from the 4/mmm group into its mmm subgroup occurs, and the symmetry of the initial tetragonal state is retained in the resulting reciprocal lattice plotted using orthorhombic components [3.20-22].

It is seen from the reciprocal lattice that the (100) reflection must be split into four components after a phase transition. The splitting has projections both along the vector of the reciprocal lattice and perpendicular to the vector. The splitting along the vector corresponds to two lattice parameters $a$ and $b$ of an orthorhombic cell along the initial tetragonal direction. The splitting perpendicular to [100] corresponds to the twinning angle $\rho$, which is determined from $\rho = (1-a/b)$.

To confirm the twin nature of the HTSC macrostructure the twinning angle was also measured independently by angular scanning topography. A typical topogram of an angular scan of reflection from (100) planes of 1-2-3-$O_x$ single crystals is displaced in Fig.3.4a. The x-coordinate corresponds to the interplanar distance, and the **y**-coordinate is the misorientation angle of crystal fragments around the normal to the crystal plane. In the samples investigated the normal was coaxial with the [100] direction. In accordance with the scheme in Fig.3.3d there are four types of spots seen on the topogram. The spots A and A′ correspond to the reflection of one twin complex by the planes a and b. The spots B and B′ are due to reflection of another twin complex whose twinning plane is perpendicular to the first one. The angle between the spots A and A′ along the **y**-coordinate equals 0.9° ±0.05°. The estimated value of the twinning angle from the lattice parameters is 0.87°.

For further confirmation of the twin character of the 1-2-3-$O_x$ substructure, topograms were measured for reflection by (110) and (001) planes. The topographic image of reflections from (001) was not split. The topogram obtained for reflection from the {110} planes is shown in Fig.

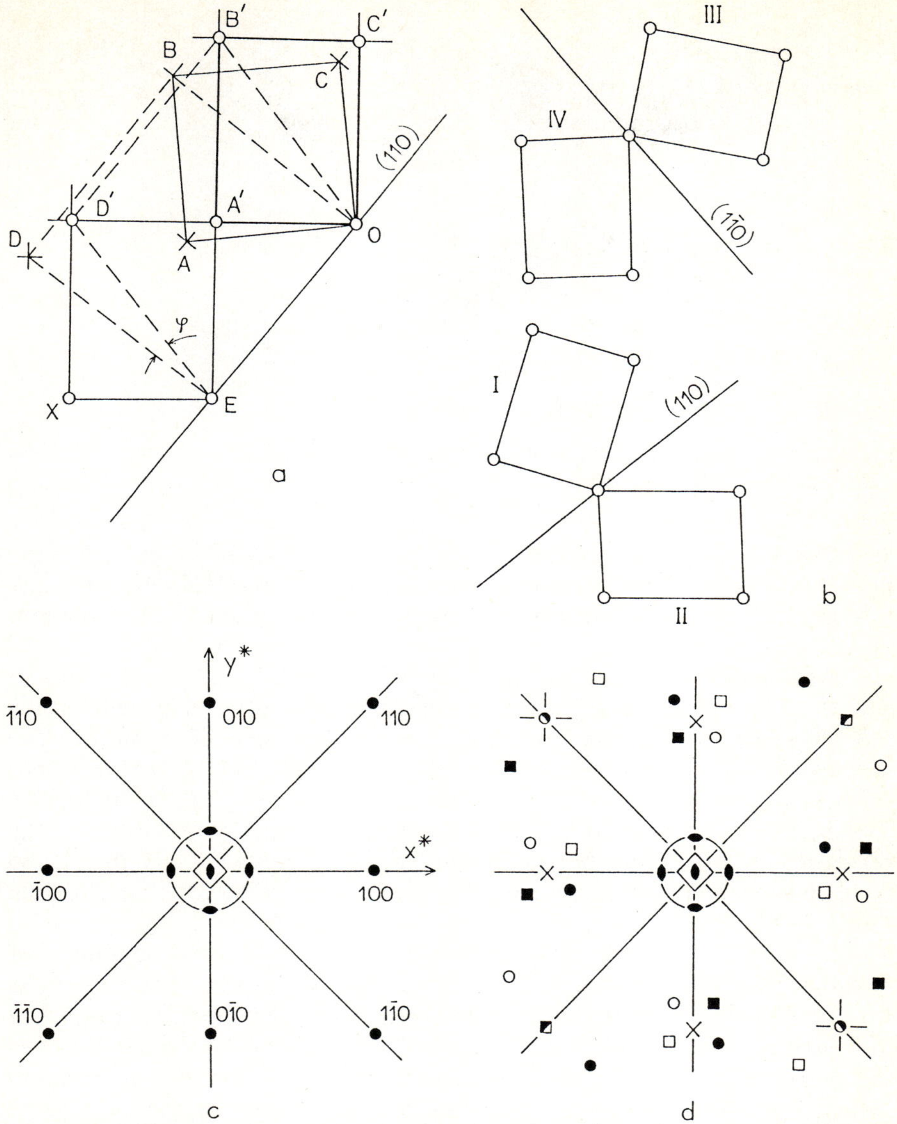


**Fig.3.3a-d.** Twinning of a tetragonal lattice along the {110}/⟨110⟩ system: (**a**) Twinning scheme, tilting a square to make a rectangle. (**b**) Resulting four orientation states. (**c**) Fragment of the reciprocal lattice of the original phase. (**d**) Superposition of the reciprocal lattice of four orientation states of the orthorhombic phase; × marks the positions of a lattice site of the tetragonal phase

3.4b. The image coincides with that of the reciprocal lattice point (110), see the scheme in Fig.3.3d. The central spot B + B′ represents the reflection from the (110) planes parallel to the twinning plane (Fig.3.3a). The spots A and A′ are the reflections from (110) planes perpendicular to the twinning plane (Fig.3.3b). The angle between A and A′ equals the twinning angle, in

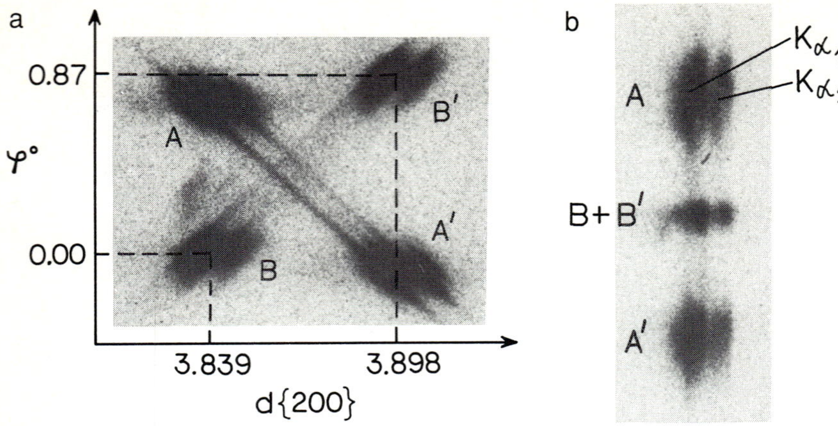


**Fig.3.4.** Angular scanning topogram of a twinned $GdBa_2Cu_3O_{7-\delta}$ crystal: (**a**) {200} reflection; (**b**) {110} reflection

accordance with the scheme in Fig.3.3. The data obtained completely support the scheme proposed for the transformation of an initial tetragonal lattice into an orthorhombic one by shearing only along the ⟨110⟩ directions [3.23,24].

Analogous investigations were carried out for $La_2CuO_4$ single crystals and the following characteristics of the twinning were established: then {100} are the twin planes, and ⟨100⟩ are the directions of twinning. The schemes in Fig.3.3 fully agree with the case of $La_2CuO_4$ if the directions (planes) of the type ⟨110⟩/{110) are substituted by ⟨100⟩/{100) correspondingly. In contrast to 1-2-3-$O_7$ systems, we have never observed the formation of incoherent boundaries in $La_2CuO_4$ crystals. Only the A and A′ spots were noted on topograms taken under the same conditions as that shown in Fig.3.4a.

Initially it had been assumed that the angle between twin planes of neighbouring complexes must be equal to 90°. However, it can be seen in Fig.3.4a that the distance between the A and B reflections is not equal to that between the A′ and B′ reflections. This means that the angle between the corresponding (100) or (010) planes in neighbouring regions is different from 90°. This is confirmed by topograms of crystals containing a large number of twin complexes. An example of such a crystal topogram for $GdBa_2Cu_3O_{7-\delta}$ is displayed in Fig.3.5. This result is also supported by the electron-microscopic investigations. Figure 3.6 depicts the diffraction images of two parts of $GdBa_2Cu_3O_{7-\delta}$ crystal. One of the parts has the conjunction boundary parallel to the twin plane (Fig.3.6a), and the other has the boundary tilted at the twin planes. Figure 3.7 exhibits the microdiffraction pattern corresponding to these parts. It is seen that for the case in Fig.3.6a every reflection on the microdiffraction pattern is split symmetrically in accordance with the twin-planes orientation. The diffraction spots for the crystal in Fig.3.6b are spaced asymmetrically with respect to the vector of the reciprocal lattice, analogously to what was obtained on the

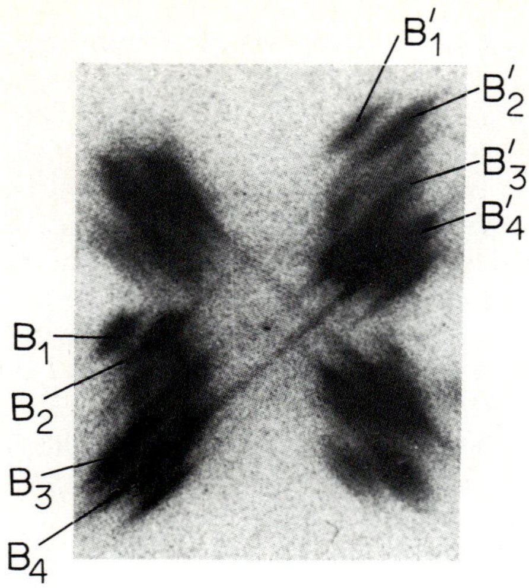


**Fig.3.5.** Angular scanning topogram of a polydomain $GdBa_2Cu_3O_{7-\delta}$ crystal. The separations between reflection spots are not all the same, indicating that the angle between twin zones is not always 90°

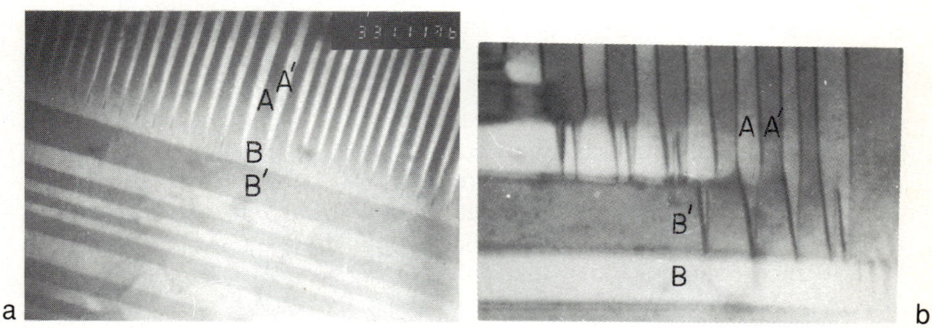


**Fig.3.6.** Electron-microscopy diffraction contrast patterns of the twinning fragments of a $GdBa_2Cu_3O_{7-\delta}$ crystal. (**a**) Boundary parallel to twin plane; (**b**) boundary tilted to twin plane

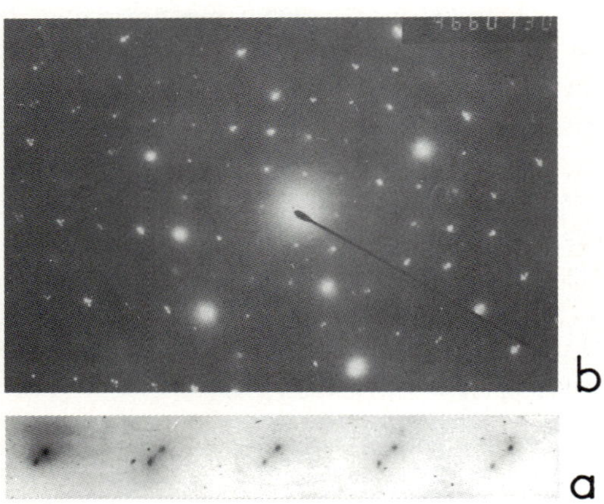


**Fig.3.7.** Electron-microdiffraction patterns of the fragments in Fig.3.6

angular-scanning X-ray topogram. Comparing the results of electron microscopy with the X-ray study, we can conclude that when the incoherent boundary does not coincide with the twin plane, the angle between the twin planes in neighboring twin complexes is not 90° [3.25].

The deviation of the angle between twin complexes from 90° might be due to the presence of an excessive number of one-sign dislocations inside the conjunction boundary. However, the electron-microscopy investigations do not confirm this. The existence of additional misorientation can be explained by assuming a continuous change of a crystallographic parameter inside the twin boundary. Such changes are shown on the topogram in Fig. 3.4.

It is seen in that figure that reflections of the type A (B) are linked with the A′ (B′) reflections by diffuse threads. These threads or "bars" show the twin boundaries to be transitional regions along which the parameter a of one twin orientation continuously transforms into the parameter b of another twin orientation, and vice versa. The change of the lattice parameters along such a region is accompanied by a simultaneous change in the tilt angle of the crystal planes from one twin orientation A (B), for instance, into another twin orientation A′ (B′). Figure 3.8 shows the topogram of a $YBa_2Cu_3O_7$ crystal part consisting of only one twin complex. The line AA′ indicates a coherent twin boundary structure. The scheme of such a twin-boundary structure is displayed in Fig. 3.9.

In addition to the transitional regions associated with coherent twin boundaries mentioned above, there exist transitional zones at the junction of twin complexes with perpendicular twin planes. In the topograms in Fig.3.10 such regions are depicted by the weak bars of reflection AB, A′B, BA′ and AB′. The lines AB′ and BA′ (visible only in the films) characterize continuous changes of lattice parameters from $a$ to $b$ along the transitional region but with a conservation of the orientation of the reflecting planes. The lines AB and A′B′ characterize the transitional regions along which only the orientation of crystal planes changes but the interplanar distances remain the same. The schemes of such transitional regions are presented in Fig. 3.11. Figure 3.11a indicates an incoherent boundary between complexes with perpendicular twin planes when the direction of the shift of the initial tetragonal cell is directed out of the twin boundary. Figure 3.11b illustrates

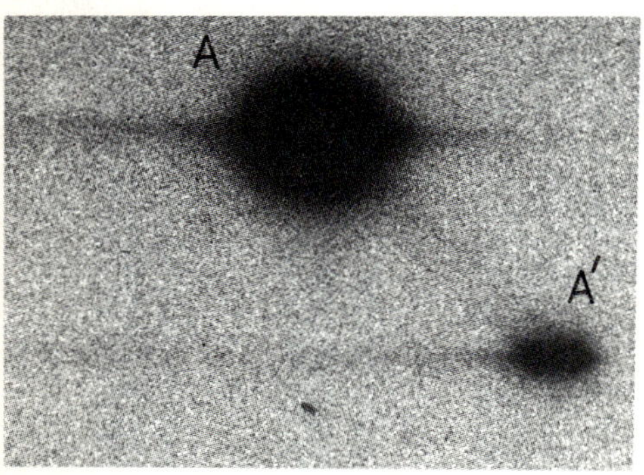


**Fig.3.8.** Angular scanning topogram of a single twin complex of $YBa_2Cu_3O_{7-\delta}$ crystal

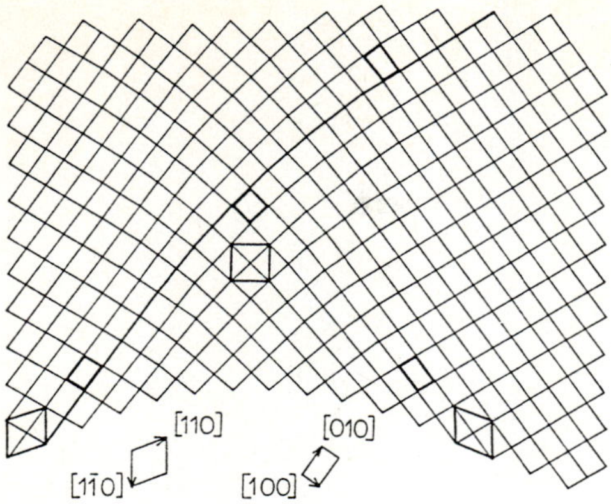


**Fig.3.9.** Scheme of a coherent twin boundary

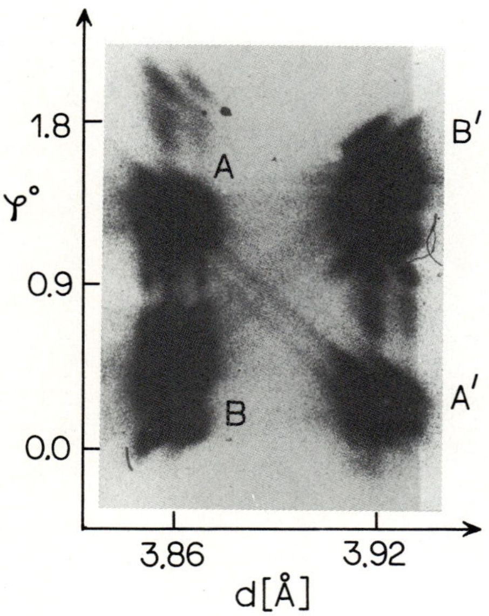


**Fig.3.10.** Angular scanning topogram of the polydomain twin complexes of a $YBa_2Cu_3O_{7-\delta}$ crystal

almost the same but with the direction of the shift toward the twin boundary. Figure 3.11c exhibits the incoherent boundaries inside a separate (single) twin orientation occurring due to the conjunction of domains with opposite directions of the shear in the initial tetragonal cells parallel to the twin boundary in different parts of the sample. Figure 3.11d depicts almost the same but with the direction of the shear perpendicular to the twin boundary. Note that along the line AA the cells retain a tetragonal symmetry.

Transitional twin boundaries are also seen in electron-microscopy images. Figure 3.12 depicts a magnified image of a $GdBa_2Cu_3O_7$ crystal reflection (620). It can be seen that the twin components are also linked to each other as is indicated in the angular scanning X-ray topograms. The AA′ line corresponds to transition regions between neighboring twins of one twin complex and indicates a continuous change of parameters and the

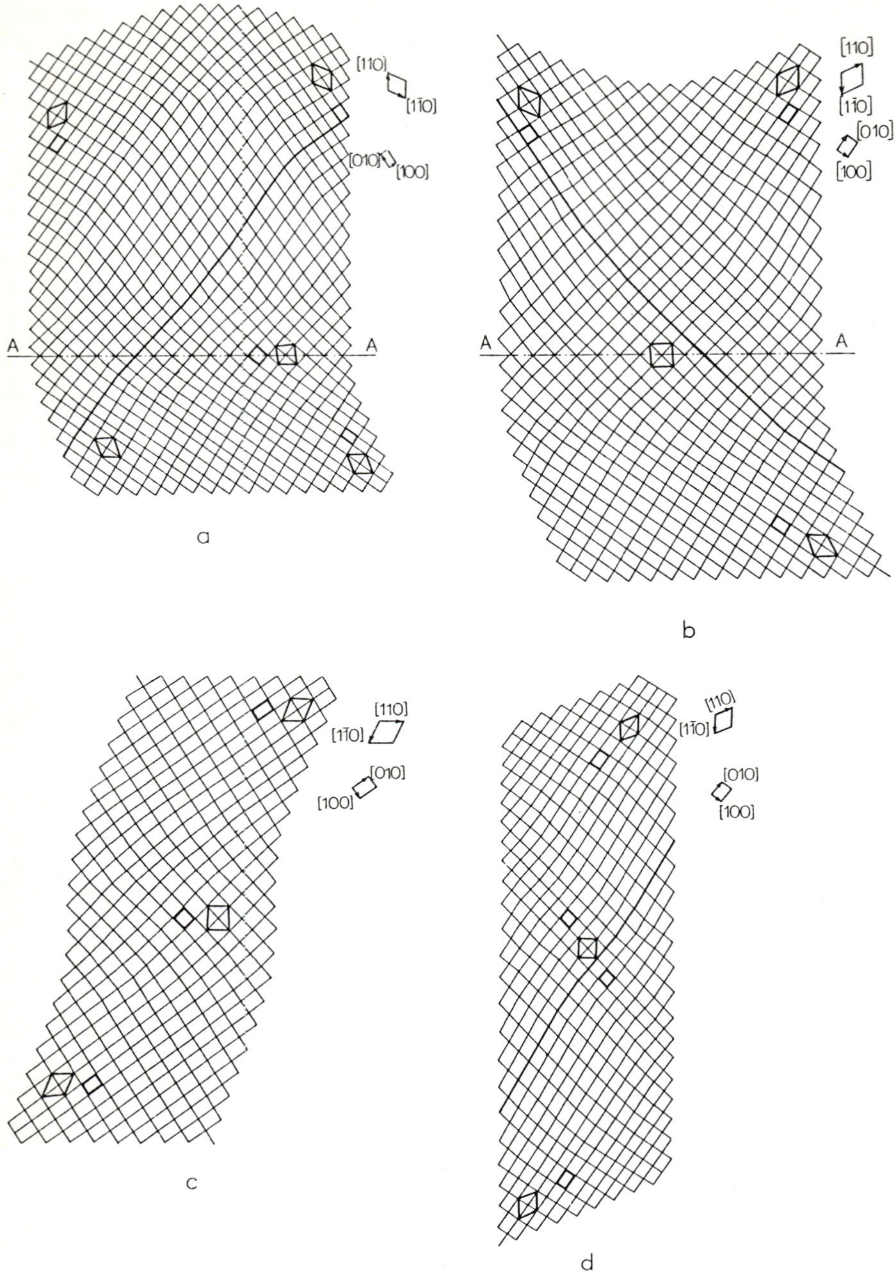


**Fig.3.11a–d.** Schemes of incoherent twin boundaries. For details see text

orientation angle of crystal planes in the transition from one twin orientation (A) to the other (A′). The BA and A′B′ lines present the transitional regions between the twin complexes with perpendicular twin planes.

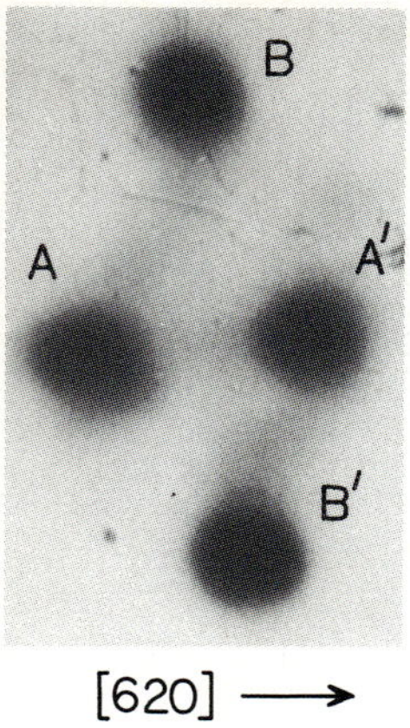


**Fig.3.12.** Microdiffraction image of the spot (620) of a twinned $GdBa_2Cu_3O_7$ crystal with incoherent boundary

It is interesting that there are no AB′, BA′ or BB′ lines. This asymmetry in diffraction is also observed in the electron-microscopy image. This may be due to the two types of junctions between twins shown in Fig.3.6. The junction between regions A and B is characterized by a smooth change of contrast in the electron diffraction image in the transition from one twin orientation to the other. The junction of the type AB′ shows a sharp change in contrast. The dimensions of the smooth transition regions AB and A′B′ are comparable to the thickness of the twin layers ($0.2 \div 0.6 \mu m$). The inhomogeneity of the contrast in the region of the junctions AB′ and A′B is estimated to be $10^{-2}$ $\mu m$ wide. The junctions of AB′ and BA′ are small and cannot be seen on a microdiffraction image. The absence of the BB′ lines may be explained by the orientation asymmetry of the twin plane with respect to the direction of the diffracted beam. The structure of the transition region at a junction of twins with perpendicular twin planes as shown in Fig.3.11, is consistent with the results of electron-microscopy observations.

To confirm the dimensions of the twin boundaries, the integral intensities of X-ray scattering were measured along the diffuse lines. The minimum peak intensity for points corresponding to intermediate values of interplanar spacing was 3% of the peak intensity of the twin components. The widths of the reflection peaks indicate that the transition regions occupy not less than 4% of the total sample volume where the signal due to transition zones between coherent twins is separated from that coming from between twin complexes. Measurement of the integral intensity for a single twin complex (as shown in Fig.3.8) yielded that the "coherent" twin boundaries occupy less than 0.5% of the total complex volume. Thus, the main part of the integral intensity detected from transition zones is determined by the boundaries between the twin complexes. Hence, the incoherent boundary thickness is equal to $2 \div 3$ $\mu m$, which is obtained from the mean dimension of the twin complexes ($100 \mu m$) and the average volume occupied by transition regions ($4 \div 6\%$). This is in agreement to within the order of magnitude with electron-microscopy measurements. It is not possible to find the dimension of the coherent twin boundary from these data.

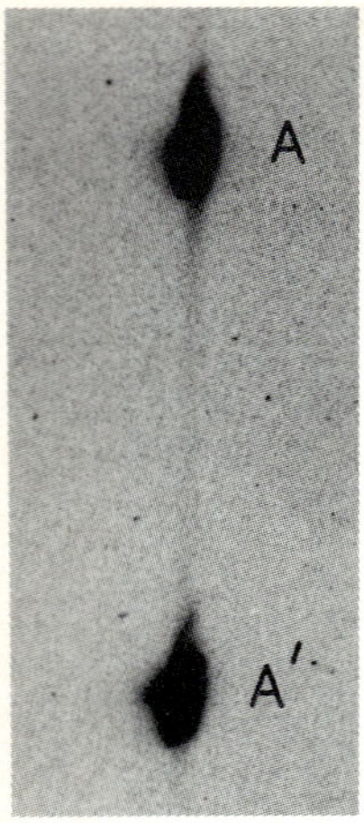


**Fig.3.13.** Angular scanning topogram of the $YBa_2Cu_3O_7$ crystal (229) reflection

What is the nature of twin boundaries in the transition zones? The presence of a tetragonal lattice state inside a transitional zone allows us to assume a change in the oxygen concentration so that in the tetragonal phase ($x< 6.5$) and in the region of a twin x increases to 7.0. This hypothesis was checked by us by looking at the topographic image of the reflection from the (229) planes. According to [3.26] the change in the parameter $\Delta c$ from 11.85 to 11.65 corresponds to a change in the composition from $x = 6.0$ to $x = 7.0$. On the topogram shown in Fig.3.13, the x-axis corresponds to the change in c and the y-axis, as in Figs.3.4,5,8, to the change of the misorientation angle between twin components, determined by the reflection from the planes (110). The rod between the spots A and A′ indicates the structure of the transition zones between these twin orientations. Because the rod does not bend along the x-axis along the transition zone no change of the lattice parameter c is observed. Therefore the oxygen composition inside the transition zone is constant. Thus, the nature of transition zones is determined by oxygen reordering along the *a* and *b* directions, but not by an oxygen concentration change, as might have been assumed.

Analogous investigations were also performed for $La_2CuO_4$. However, no rods between single-twin components could be observed. This means that either there are no transition zones in these crystals or that they are so small that the reflection intensity from them is too low.

## 3.4 Quasi-Twins

This data on the nature of twin boundaries are supported by experiments on $HoBa_2Cu_3O_7$ [3.27]. Crystals were produced by the following method:

In the first stage, the ceramics were synthesized, which included mixing the initial oxides $Ho_2O_3 \cdot 8BaCo_3 \cdot 15CuO$, heating for one hour in air up to $T = 920°C$, holding the sample at this temperature for 24 hours and then cooling slowly in the furnace. In the second stage, a single crystal was produced by the following steps:

- heating of the ceramics from 20° up to 950° C at a rate of 320° per hour, from 950° to 1050° C at a rate of 90° per hour;
- exposure at 1050° C for 6 hours;
- cooling from 1050° C to 1000° C at a rate of 4° C per hour, from 1000° to 950° C at a rate of 90° C per hour, from 950° to 700° C at 20° per hour and then the furnace was turned off.

For the X-ray investigations, crystal samples being 50 $\mu$m thick were selected. Optical investigations of the samples in polarized light did not reveal a developed twin structure, but confirmed the presence of an orthorhombic phase. Thermal measurements of the magnetic susceptibility suggested a superconducting transition in the range of 80÷40 K.

The X-ray diffractometry and X-ray topography measurements of these crystals were also carried out. The size of the crystal region studied varied greatly; the minimum size was 50×5 $\mu m^2$. Figure 3.14 shows typical topograms of the angular scans obtained from two regions of the same sample (50×5.0$\mu m^2$ in size). Comparison with the topograms of $YBa_2Cu_3O_7$ and $GdBa_2Cu_3O_7$ crystals demonstrates the absence of pronounced twin states with well-defined lattice parameters and twin angle. The crystal is only a set of transition zones with a structure similar to that of twin boundaries. In contrast to twin boundaries, however, the sizes of such zones are several and even tens of micrometers. The topographic patterns of the transition zones in $HoBa_2Cu_3O_7$ are similar to the diffuse lines (the bars) of the type AA′ and BB′ for twin crystals. The change of lattice parameters from $a_{min}$ into $a_{max}$ occurs along these rods. The tetragonal

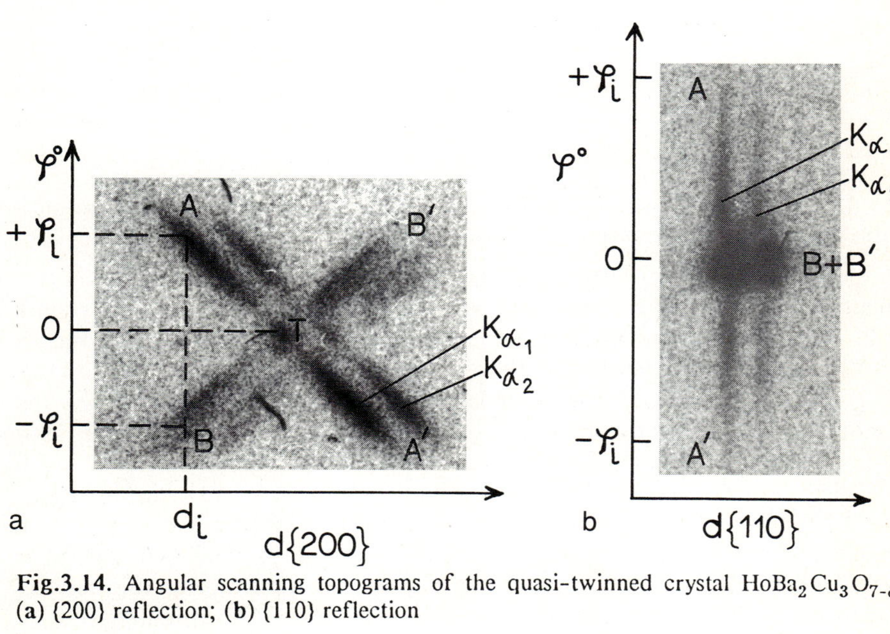


**Fig.3.14.** Angular scanning topograms of the quasi-twinned crystal $HoBa_2Cu_3O_{7-\delta}$: (**a**) {200} reflection; (**b**) {110} reflection

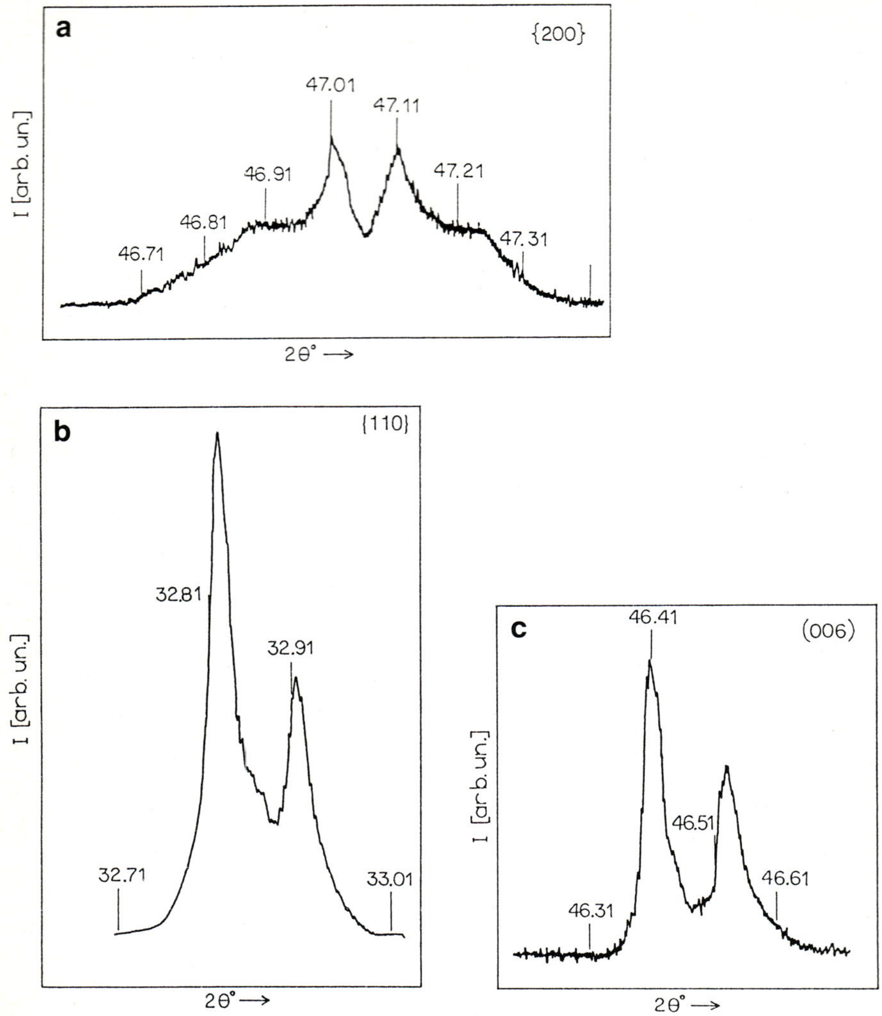


**Fig.3.15.** The diffractograms of the quasi-twinned $HoBa_2Cu_3O_{7-\delta}$ crystal: (**a**) {200} reflection; (**b**) {110} reflection; (**c**) (006) reflection

phase state corresponds to the intersection of the lines AA′ and BB′ when a = b. The change in the lattice parameters along the transition zones is shown on diffractometry patterns for the reflection from (100) planes (Fig.3.14a). The lines AA′ and BB′ correspond to crystal fragments inside which the tilts of the reflecting planes with a well-defined interplanar distance {200} are opposite. Such crystal fragments are called *quasi-twins*.

The term "quasi-twin" is suitable since twin planes and twin directions can be defined for such formations. The {110} planes are also the twin planes and the twin directions coincide with the ⟨110⟩. The twin character of such formations is supported by the fact that every orthorhombic cell inside a quasi-twin may be assumed to be in the form of an initially tetragonal cell, deformed in the [110] direction for the line AA′ and [$\bar{1}$10] for BB′. Then, the tilt angle of the (100) planes of any orthorhombic cell will be defined by an orthorhombicity parameter a/b and will coincide with half the twin angle $\rho = 1 - a/b$. The angles $\rho$ measured from the topogram in Fig.3.14a are in good agreement with calculated values obtained for the parameters a along the lines AA′ and BB′.

The twin planes {110} do not undergo changes in lattice parameters for either quasi-twins or simple crystals. Figure 3.14b shows the topogram of the reflection from the {110} planes of a quasi-twin crystal. According to the above statement the B + B′ spot corresponds to reflections from the (110) planes parallel to the twin plane. The rods BA and B′A′ relate to reflections from the (1$\bar{1}$0) planes perpendicular to twin planes (Fig.3.14a). In contrast to the different orientation states of a conventional twin crystal the (110) planes for a quasi-twin undergo a bending across the boundaries. The topogram also reveals that the {110} parameters are constant for the (110) planes as well as for the (1$\bar{1}$0) ones perpendicular to them. This result is confirmed by the diffraction patterns (Fig.3.15), where the separation of the $K_{\alpha_1}$-$K_{\alpha_2}$ doublet can be seen even at small angles of diffraction. This supports the assumption proposed above that the formation of orthorhombic cells inside a quasi-twin is due to the shear of initial tetragonal cells in the ⟨100⟩ directions.

Thus quasi-twins are analogous to conventional twins, in which a continuous change of lattice periods and orientation of crystal planes occur symmetrically about a tetragonal phase layer. A scheme of a quasi-twin in accordance with the results of the diffraction experiment, is illustrated in Fig.3.16. The orthorhombicity parameter change can be related to two structural processes: a change in oxygen concentration in the crystal between $0_6$ and $O_7$, and a change in the degree of oxygen ordering in the basal a-b plane with constant oxygen concentration. Our measurements demonstrate that the parameter c for a quasi-twin crystal remains constant to within $\pm 3 \cdot 10^{-4}$ Å over the entire crystal volume. Thus, it can be concluded that an orthorhombicity parameter change inside a quasi-twin, as in the case for twin boundaries, occurs due to oxygen ordering in the a-b basal plane, at constant O concentration. With this interpretation it is possi-

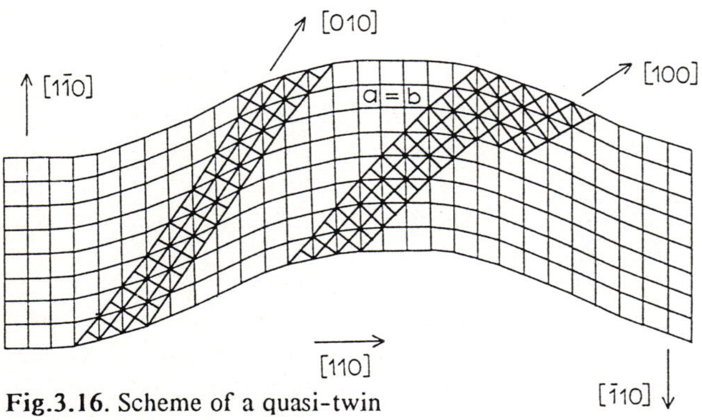


**Fig.3.16.** Scheme of a quasi-twin

ble to understand the broad temperature width of the superconducting transition in quasi-twins. It is determined by the dependence of $T_c$ on the degree of orthorhombicity or, which is the same thing, on the degree of oxygen ordering in the basal plane which can vary greatly inside one quasi-twin.

## 3.5 Twins in Epitaxial 1-2-3-$O_x$ Films on Tetragonal Substrates

Epitaxial films based on the 1-2-3 system possess abnormally high critical currents and have attracted considerable attention due to the interrelation of the structure and the physical properties. We have investigated analogous epitaxial films of $YBa_2Cu_3O_7$ on tetragonal substrates aiming to study the possible abnormalities of twin structures at the following junctions: twin film-tetragonal substrate, twin substrate-tetragonal film, and twin sub-

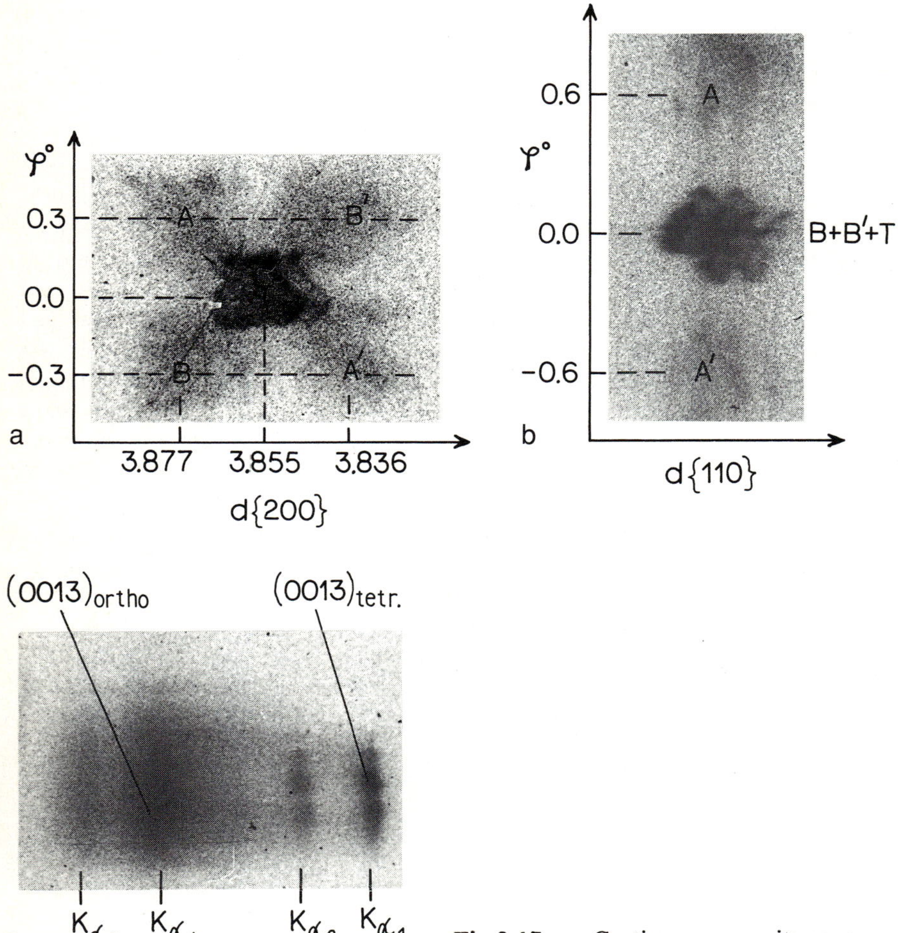


Fig.3.17a–c. Caption see opposite page

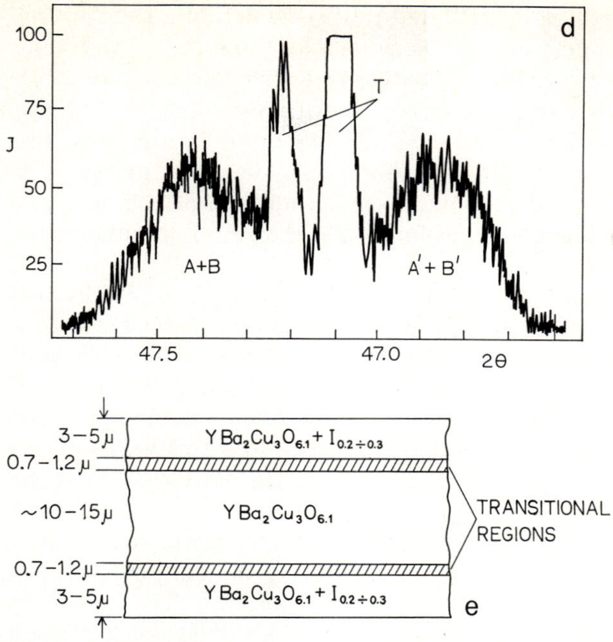


**Fig.3.17.** Iodinated single crystals of $YBa_2Cu_3O_x$. (**a**) Angular scanning topogram of a 1-2-3+1 single crystal. A and A′ are the images of twin complexes with twinning planes (110) in reflections from the (200) (A) and (020) (A′) planes. B and B′ are the images of twin complexes with twinning planes (1$\bar{1}$0) in reflections from (200) (B) and (020) (B′) planes. T stands for the image of the tetragonal phase. (**b**) Angular scanning topogram of the (0012) reflections. (**c**) Angular scanning toppogram of the (0013) reflection. (**d**) Diffraction intensity distribution of the {200} reflections corresponding to the topogram of part (**a**). (**e**) Scheme of the real structure of the iodinated single crystal

strate-twin film. Of these three systems we have been able to understand only one, namely, the twin structure of analogues of epitaxial films on tetragonal substrates.

We chose to examine $YBa_2Cu_3O_6$ crystals, which were annealed in iodine vapor at 400° C and under a pressure of 6 atm. It was assumed [3.28-30] that under these conditions iodine penetrates into the crystal lattice and occupies the positions of chain oxygens, which leads to orthorhombicity and superconductivity. However, according to more detailed investigations, destruction of the ceramics with formation of iodates and iodites, and liberation of atomic oxygen which is freely diffused along the lattice 1-2-3-$O_{6.5}$, leading to the composition $YBa_2Cu_3O_{6.65}$, can also occur [3.31]. The depth of penetration depends on the time and temperature of annealing. The resulting surface layer is analogous to the epitaxial film 1-2-3-$O_{x>6.5}$ conjugated coherently with the tetragonal matrix 1-2-3-$O_{x=6.0}$.

Again angular scanning topography and precision diffractometry were used to study the structure of the iodinated single crystals. Figures 3.17a-c present topograms for the {200}, {100} and (0013) reflections. The splitting of the {200} reflections into four components and of the {110} reflections into three components indicates the formation of a twinned structure (spots A, A′ and B, B′) while the tetragonal matrix does persist (spot T). The appearance of all possible twin orientations indicates the presence of a great number of twin components with perpendicular twinning planes. The identical appearance of the topographic image for all twin orientations at a linear resolution of 5 $\mu$m leads to an estimate of the sizes of the twin complexes of tens of micrometers.

The diffraction pattern corresponding to the topogram of Fig.3.17a is shown in Fig.3.17d. It is seen that the line widths of the iodinated orthorhombic phase are larger than that of the tetragonal matrix. Comparison of the half-width for difference-order reflections suggests that the width of the (200) and (020) lines depends not on the uniformity of the iodine depth distribution in the twinned crystal fraction, but on the small size of the twins. From the half-width of the peaks one can estimate the mean width of the twins as 150÷200 Å.

The transition region between the initial tetragonal matrix and the surface layer of the orthorhombic phase formed during the annealing is seen on the topogram scanned from the *c*-plane (Fig.3.17c). The transition region is a diffuse background between the reflections from the tetragonal matrix and the orthorhombic layer. From the reflection-intensity distribution and the crystal thickness, and by assuming that the X-ray absorption is linear, one can estimate the sizes of the regions of the orthorhombic and tetragonal phases and of the transition layer: 3÷5 $\mu$m and 0.7÷1.2 $\mu$m, respectively (Fig.3.17e).

More precise measurements yield the following lattice parameters: a = b = 3.855 Å and c = 11.822 Å for the tetragonal matrix, and a = 3.836 Å, b = 3.877 Å and c = 11.722 Å for the iodinated layers of the orthorhombic phase. Based on the dependence of the lattice constants on the oxygen content, one can estimate that c = 11.722 Å corresponds to a hypothetical oxygen content of x = 6.65.

These data permit us to draw certain conclusions about the twin structure of epitaxial films of 1-2-3-$O_7$ on single-crystal substrates. If the (100) plane of a strontium-titanate crystal is used as the substrate, then a continuous transition from the substrate tetragonal lattice into an orthorhombic one on the surface corresponds to coherent conjugation of the film and substrate. The necessity of conjugation of the orthorhombic lattice of a 1-2-3-$O_7$ film in the *a* and *b* directions with the tetragonal matrix must lead to splitting of the film into twins of small sizes. That the twins must be small can be understood when one considers the competition between the energy of twin boundaries and the elastic energy determined by the deformation of the a-c (b-c) planes which bend with increasing twin size along the substrate (Fig.3.18). According to this scheme, the angle $\varphi$ of the deformation bending of the a-c planes is proportional to the twin size, and inversely proportional to the transition-layer thickness. This type of structure does not have long twin boundaries and is likely to manifest strong pinning

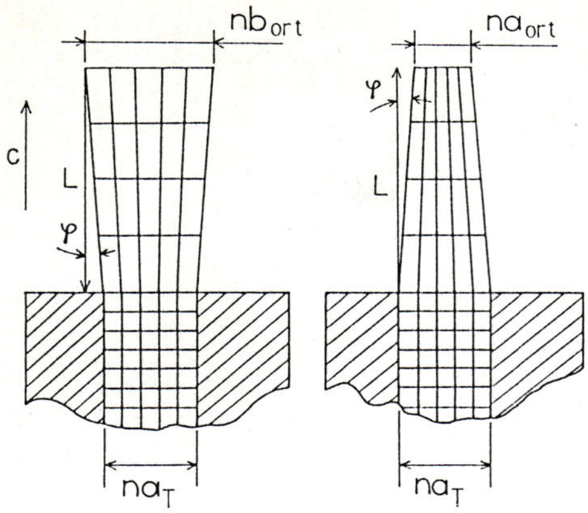


**Fig.3.18.** Scheme of the coherent conjugation of a tetragonal substrate with the twin orientation of an orthorhombic epitaxial film

properties. The latter may point to a reason for the increase in critical currents in real epitaxial films on 1-2-3-$O_x$ compounds.

The date obtained also suggest the structure of the interphase boundary which is the transition region between the surface layer of the orthorhombic phase and the tetragonal matrix. The topographic images of the reflections from the (229) plane were used for this (Fig.3.19). As in Fig.3.13, the x-axis is directed along the change in the c parameter, the y-direction represents the angular misorientations of twin states. It can be seen in Figs.3.17b and 19 that along with the twin components A,A′ and B,B′ from the orthorhombic phase on the surface, there are images of the tetragonal matrix T and the interphase boundary. The latter is seen in the topogram in the form of a rod among the reflections B + B′ and T, and in the form of rods among the reflections A′, A and T. The image may be interpreted in the following manner: a given misorientation angle of the substructure elements corresponds to each oxygen concentration along the interphase boundary (the c direction) in the (110) plane. The value of this angle inside the layer of the orthorhombic phase on the sample surface becomes constant and equal to the twin angle. Thus, along an interphase boundary, just as on

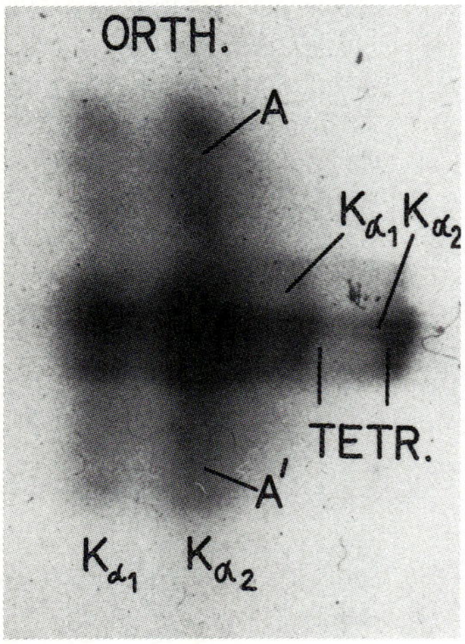


**Fig.3.19.** Angular scanning topogram of a $YBa_2Cu_3O_{6.1}$+I crystal. (229) reflection

an intertwin boundary, a continuous change of crystallographic parameters from the values typical of the substrate to crystallographic parameters of the twin orientation of an orthorhombic phase layer, occurs. The angle between twin orientations changes continuously from zero on the substrate to the twin angle of the orthorhombic layer on the surface.

## References


3.1 J.F. Nye: *Physical Properties of Crystals* (Clarendon, Oxford 1957) p.585
L.A. Shuvalov (ed.): *Modern Crystallography IV: Physical Properties of Crystals*, 2nd edn., Springer Ser. Solid-State Sci., Vol.37 (Springer, Berlin, Heidelberg 1992)
3.2 I. Zheludev, L. Shuvalov: Kristallographiya **1**, 681-688 (1956)
3.3 I. Zheludev, L. Shuvalov: Izv. Akad. Nauk SSSR, Ser. fiz. **21**, 264-274 (1957)
3.4 L. Shuvalov: Kristallographiya **8**, 617-624 (1963)
3.5 N. Afonikova, I. Shmyt'ko, V. Shekhtman: Izv. Akad. Nauk SSSR, Ser. fiz. **43**, 1611-1618 (1979)
3.6 M. Klassen-Neklyudova: *Mechanical Twinning of Crystals* (Izd. AN SSSR, Moscow 1961) p.261
3.7 T. Kontorova: J. Expt. Theor. Phys. **12**, 68-78 (1942)
3.8 V. Zhirnov: J. Expt. Theor. Phys. **35**, 1175-1180 (1958)
3.9 E. Little: Phys. Rev. **98**, 978-984 (1955)
3.10 N. Afonikova, V. Borovikov, I. Shmyt'ko: Fiz. Tverd. Tela **29**, 813-817 (1987)
3.11 L. Dorosinsky, M. Indenbom, V. Nikitenko, B. Farber: J. Expt. Theor. Phys. Lett. **49**, 156-159 (1990)
3.12 V. Kogan: Phys. Rev. Lett. **62**, 3001-3003 (1989)
3.13 J. Mannhart, P. Chaudhari, D. Dimos, C. Tsuei, T. McGuire: Phys. Rev. Lett. **61**, 2476-2479 (1988)
3.14 D. Dimos, P. Chaudhari, J. Mannhart, F. LeGones: Phys. Rev. Lett. **61**, 219-222 (1988)
3.15 P. Chaudhari, J. Mannhart, D. Dimos, C. Tsue, J. Chi, M. Oprisko, M. Shenermann: Phys. Rev. Lett. **60**, 1653-1656 (1988)
3.16 P. Kes: Physica C **153-155**, 1121-1126 (1988)
3.17 L. Vinnikov, L. Gurevich, G. Yemel'chenko, Yu. Ossipyan: Solid State Comm. **67**, 421-423 (1988)
3.18 T. Matsushita, M. Iwakuma, Y. Sudo: Jpn. J. Appl. Phys. **26**, 1524-1526 (1987)
3.19 N. Afonikova, V. Shekhtman, I. Shmytko: Fiz. Tverd. Tela **27**, 3201-3207 (1985)
3.20 Yu. Ossipyan, N. Afonikova, G. Yemel'chenko, T. Parsamjan, I. Shmyt'ko, V. Shekhtman: J. Expt. Theor. Phys. Lett. **46**, 189-192 (1987)
3.21 Yu. Ossipyan, N. Afonikova, T. Parsamjan. I. Shmyt'ko, V. Shekhtman: J. Expt. Theor. Phys. Lett. **47**, 501-504 (1988)
3.22 Yu Ossipyan, V. Shekhtman, I. Shmytko: Physica C **153-155**, 970-971 (1988)
3.23 Yu. Ossipyan, V. Shekhtman, I. Shmyt'ko: Z. Kristallographie **185**, 428 (1988)
3.24 Yu. Ossipyan, N. Afonikova, D. Batova, V. Goncharov, G. Yemel'chenko, M. Indenbom,. E. Suvorov, V. Shekhtman, I. Shmytko: Fiz. Tverd. Tela **31**, 131-138 (1989)
3.25 I. Shmyt'ko, V. Shekhtman, Yu. Ossipyan, N. Afonikova: Ferroelectrics **97**, 151-170 (1989)
3.26 R. Cava, B. Battlog, C. Chen, E. Rietman, S. Zahurak, D. Werger: Phys. Rev. B **36**, 5719-5722 (1987)

3.27 Yu. Ossipyan, N. Afonikova, V. Borodin, L. Chernyshova, V. Shekhtman, I. Shmyt'ko: Fiz. Tverd. Tela **31**, 200-204 (1989)
3.28 Yu. Ossipyan, O. Zharikov, G. Logvenov, N. Sidorov, V. Kulakov, I. Shmyt'ko, I. Bdikin, A. Gromov: Physica C **165**, 107-110 (1990)
3.29 Yu. Ossipyan, O. Zharikov, N. Sidorov, V. Kulakov, D. Mogilyansky, R. Nikolaev, V. Shekhtman, O. Valegova, I. Romanenko: J. Expt. Theor. Phys. Lett. **48**, 225-227 (1988)
3.30 Yu. Ossipyan, O. Zharikov, G. Novikov, N. Sidorov, V. Kulakov, L. Sypavina, R. Nikolaev, A. Gromov: Physica C **159**, 137-140 (1989)
3.31 O. Mysochko, Yu. Ossipyan, O. Zharikov, R. Nikolaev, N. Sidorov, V. Kulakov, A. Gromov: Sverkhprovodimost (Sov. Superconductivity) **4**, 954-956 (1991)

# 4. Deformation, Structure and Properties of High-$T_c$ Superconducting Ceramics and Single Crystals

V.S. Bobrov

After the discovery of high-$T_c$ superconductivity by *Bednorz* and *Müller* [4.1] a great number of publications on studies of this astonishing physical phenomenon appeared. Currently, several High-$T_c$ SuperConductors (HTSCs) are known, and their electrical, magnetic, optical and other physical properties are being intensively studied [4.2]. In this chapter we shall discuss the results for the mechanical properties of HTSCs, namely, strength, plasticity and microhardness.

Although these properties cannot be classified together with fundamental characteristics such as electrical and magnetic properties, it should be borne in mind that they are important in studies of the influence of deformation (which can be necessary in preparing HTSCs for applications) on the structure and properties of HTSCs. Of particular interest are combined studies of mechanical, structural and superconducting characteristics. We hope that our results and the data given in the other chapters will enable the reader to gain some insight into the problems posed by the novel superconductors. However, we would like to warn the reader that studies in this field are far from complete; we are witnessing only the first steps in this direction.

In this chapter we shall pay special attention to the results obtained for $YBa_2Cu_3O_{7-x}$, i.e., those compounds which have until now been studied most thoroughly [4.3]. In the example of Y-Ba-Cu-O, we shall show that not only electrical but also the mechanical properties alter dramatically with varying content and position of the oxygen in the crystal lattice. Thus we shall emphasize again its particular role in determining different physical properties of high-$T_c$ superconductors. The studies of deformation in other HTSC compounds will be discussed only briefly. We shall only touch upon the results of internal friction. These investigations require special consideration which is beyond the scope of this chapter.

## 4.1 Problems of Deformation of High-$T_c$ Superconductors

At low and moderate temperatures, single crystals and ceramics of the known high-$T_c$ compound superconductors are brittle [4.4-10]. This creates certain difficulties in their investigation. Other problems arising in studies of mechanical properties of single crystals are related to the shapes and sizes of samples (thin pellets) and the occurrence of microcracks [4.4-6, 9b]. Therefore, investigations of deformation processes in this temperature range are mainly performed by a microindentation technique [4.4-7]. We should also mention here the experiments by *Peschanskaya* et al. [4.8] on inelastic microdeformation due to low-temperature creep of Y-Ba-Cu-O ceramics and the data on the critical breaking stress [4.9, 10].

The plasticity of HTSC ceramics increases at elevated temperatures [4.9, 11-13]. In this case conventional methods may be employed to investigate stress-strain curves [4.14]. At high temperatures, however, other problems arise, which are related to phase-composition stability and, specifically, to sample-composition stability with respect to oxygen [4.9, 15]. The problems associated with single-crystal deformation remain, too. At present, only a few works are available in which Y-Ba-Cu-O single crystals were deformed by bending at elevated temperature, e.g., [4.9b]. Because of technical difficulties, practically no data exist on microindentation of ceramics and single crystals in the high-temperature range [4.5]. This creates a gap between the results obtained at high and low temperature ranges.

## 4.2 Deformation of Y-Ba-Cu-O Ceramics and Single Crystals

Methods of dynamical deformation (constant-rate loading) and creep (constant load) are those usually employed to study mechanical properties of materials [4.14, 16, 17]. The results obtained for HTSC compounds under dynamical deformation and creep will be considered in this section. First, we shall briefly consider the problems of brittle fracture [4.9, 10, 14, 18] and discuss the low-temperature microcreep data [4.8, 19-22]. Then, we shall go over to the results of high-temperature deformation studies on Y-Ba-Cu-O ceramics [4.9, 11-13, 23, 24] and report on attempts to deform Y-Ba-Cu-O single crystals [4.9]. We shall present the data indicating the dependence of strength and plasticity on the oxygen concentration and distribution [4.9, 13, 23]. This section will be concluded with a brief analysis of the data on structural modifications upon straining HTSC ceramics and single crystals [4.9, 24-27].

As mentioned, we shall focus our attention on the data for Y-Ba-Cu-O compounds. The deformation of other HTSC compounds was studied less intensively. However, the available data enable one to conclude that the deformation behavior of known HTSC compounds is quite similar. Naturally, there exist quantitative differences in parameters characterizing the mechanical properties [4.9-13, 23, 24] but these differences are often related to synthesis conditions and sample-preparation technology. Therefore, their detailed discussion at this stage is not so informative. Only qualitative singularities in deformations of this class of perovskites have to be studied.

### 4.2.1 Brittle Fracture and Microplasticity

All the known HTSC compounds are brittle. In the range of low and moderate temperatures their deformation to the brittle fracture proceeds in elastic or, to be more precise, quasielastic fashion (Fig. 4.1, curve *1*) [4.4-10].

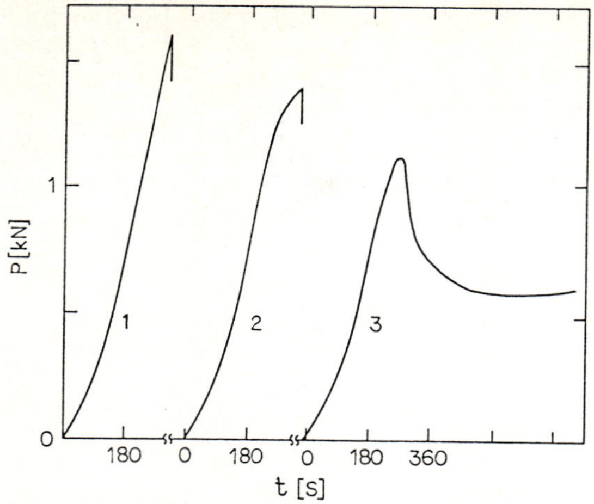


**Fig.4.1.** Deformation of $YBa_2Cu_3O_{7-x}$ ceramics under compression in a helium atmosphere at a rate of 100 μm/min: T ≃ 990 (*1*), 1030 (*2*) and 1090 K (*3*) (P denotes loading and t the time)

The transition to the plastic flow occurs only after heating the ceramic samples above a particular temperature $T_0$ [4.9]. This behavior is illustrated in Fig.4.2. It is typical not only for HTSCs, but for many other materials, including ceramics [4.14, 18]

The main parameter measurable upon brittle fracture is the ultimate strength (critical breaking stress) [4.14, 18]. In the case of ceramics this parameter is mainly determined by conditions of sample compacting and sintering rather than by intrinsic properties of the constituent crystallites. In single crystals the ultimate strength values are affected by singularities of the sample shape and their tendency to crack formation [4.9]. For this reason microindentation is so widely employed in deformation tests of HTSC single crystals and ceramic crystallites [4.4-7]. The results obtained

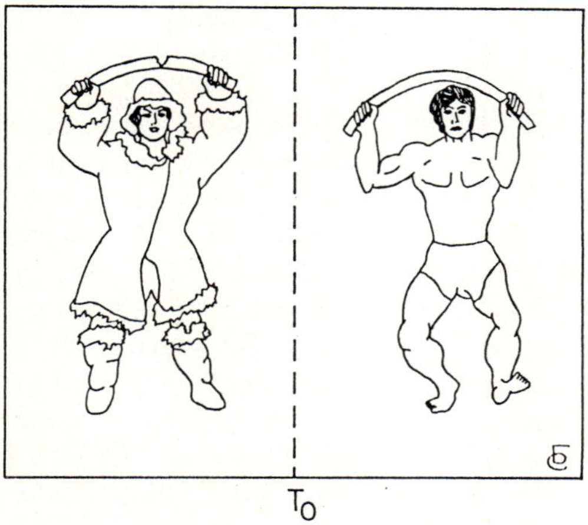


**Fig.4.2.** As the temperature is increased, the brittle fracture-to-plastic deformation transition occurs in high-$T_c$ ceramics. The transition temperature $T_0$ depends on the composition, sample synthesis and compacting conditions

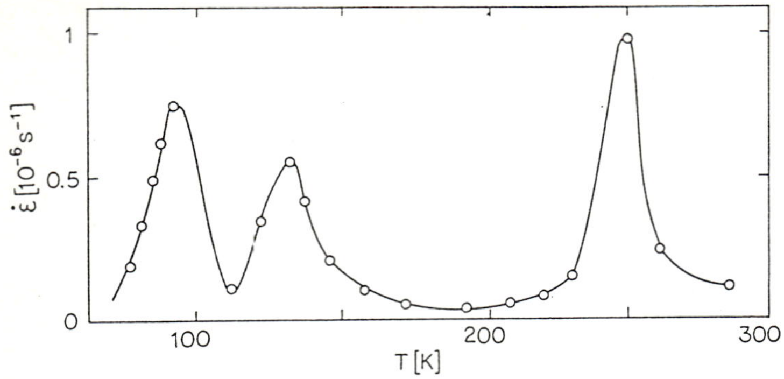


**Fig.4.3.** Spectra of inelastic deformation rates under compression of fine-grained (d $\simeq$ 1÷3$\mu$m) $YBa_2Cu_3O_{7-x}$ ceramics; $\sigma \simeq$ 10 MPa

with this technique will be considered in detail in Sect.4.4. Here we shall briefly discuss some investigations on the inelastic deformation of Y-Ba-Cu-O ceramics in the stress range below the ultimate strength [4.8, 19, 20].

We consider the two effects observed under low-temperature creep of the $YBa_2Cu_3O_{7-x}$ ceramics [4.8, 19]. We shall first discuss the temperature dependence of the rate of inelastic deformation $\dot{\varepsilon}$ and point out the occurrence of peaks on the function $\dot{\varepsilon}$(T). As an example of $\dot{\varepsilon}$(T) in the range T $\simeq$ 77÷300 K we present in Fig.4.3 the data obtained by *Peschanskaya* et al. [4.8b] for a Y-Ba-Cu-O ceramic sample. The positions of the peaks are not fixed; they depend on the ceramic's synthesis conditions and vary from sample to sample [4.8, 19, 20]. The situation here is much the same as for sound absorption in HTSC compounds, which is supported by numerous data [4.28-40]. Naturally, a certain correlation exists between these phenomena, but their simultaneous study has not been carried out so far. To date there is no unambiguous viewpoint on the nature of the peaks due to sound absorption and inelastic deformation of HTSC compounds. The role of grain boundaries, phase transitions, transformations in the oxygen sublattice, dislocation mechanisms, etc. have been discussed [4.8, 19, 20, 28-35). In fact, specific features of the temperature dependence of the sound absorption and microdeformation are apparently determined by a whole complex of factors.

Before discussing another feature of the low-temperature deformation of HTSC ceramics, we shall take a look back into history. More than 20 years ago the effect of the electronic state in classic superconductors on their dynamic deformation, creep and internal friction was discovered. Subsequently, various details of this effect, associated with a change of the electron-dislocation interaction conditions at superconducting transitions, were investigated [4.41-43]. Interest in this research still remains. It is, therefore, quite natural that *Peschanskaya* et al. [4.8a] attempted to find the analogous effect in high-$T_c$ superconductors, subjected to inelastic deformation. In their experiments, Y-Ba-Cu-O ceramics were cooled down to T $\simeq$ 77 K, and the superconductivity was destroyed by means of electric cur-

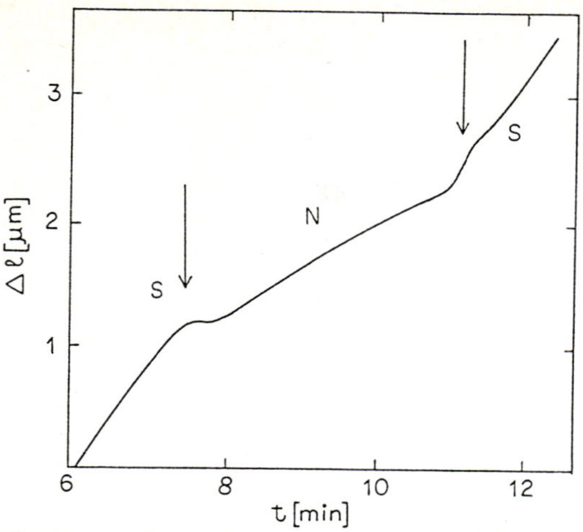


**Fig.4.4.** Portion of the creep curve for $YBaSrCu_3O_{7-x}$ ceramics ($T_c \simeq 90 \div 95K$) at $T \simeq 77$ K and $\sigma \simeq 12$ MPa, deduced from a laser interferogram. Times at which the electric current is switched on and off are shown by *arrows*, $J \simeq 25$ A/cm$^2$. The states (S: superconducting, N: normal) of the sample under deformation were controlled by a four-point potential technique

rent. The halting of superconductivity was found to be accompanied by the deceleration (or stopping) of the deformation and, vice versa, the deformation accelerated when the current was turned off (fig.4.4). This result was later confirmed in similar studies by *Soldatov* et al. [4.21], and, quite recently, *Smirnov* [4.22] reported preliminary results on the effect of a magnetic field on inelastic deformation of HTSCs. This behavior is qualitatively coincident with the above data obtained for classical superconductors. One may assume that this phenomenon in HTSCs is related to a change of the electronic state in the grain-boundary region [4.21], but more definite conclusions about its nature may be drawn only after having elucidated the mechanisms of inelastic deformation of HTSCs.

We shall not consider in detail the results of ultimate-strength studies for different HTSC compounds. Note, again, that this parameter is determined mainly by the conditions of sample preparation. The strength characteristics of Y-Ba-Cu-O ceramics may be evaluated from, e.g., the data presented in Figs.4.5,6 (Sect.4.2.2) for different groups of Y-Ba-Cu-O ceramic samples [4.9]. We shall come back again to the problems of low-temperature deformation and brittle fracture of HTSC materials in Sect.4.3, and now we will proceed with the results deduced from studies at elevated temperatures.

### 4.2.2 Plastification of Ceramics at Elevated Temperatures

Up to a certain temperature all the investigated samples of Y-Ba-Cu-O ceramics and of other HTSC compounds exhibited a low plasticity level. Prior to brittle fracture their deformation proceeds in a quasielastic fashion

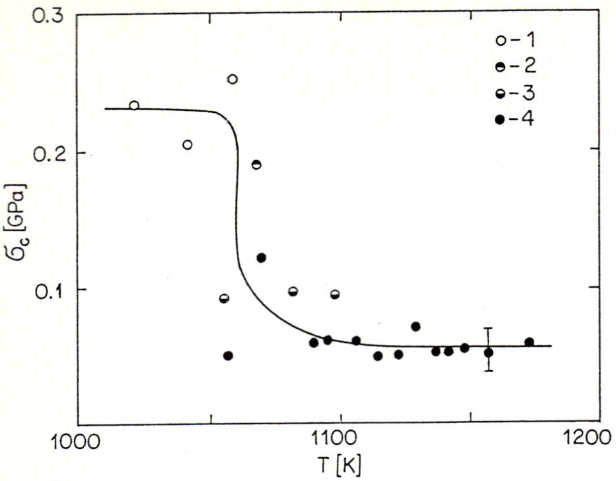


**Fig.4.5.** Temperature dependence of the brittle fracture or noticeable plastic flow stress $\sigma_c$ for typical $YBa_2Cu_3O_{7-x}$ ceramic samples: *1* brittle fracture (curve *1*, Fig.4.1), *2* and *3* different degrees of plasticity (e.g., curve *2*, Fig.4.1); *4* plastic flow (curve *3*, Fig.4.1)

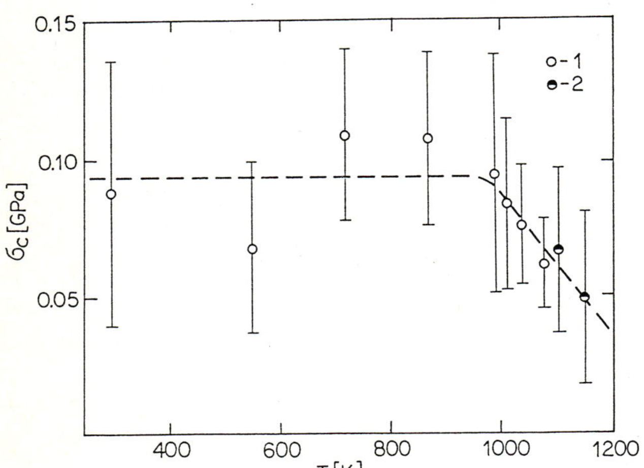


**Fig.4.6.** Temperature dependence of the ultimate strength $\sigma_c$ for $YBa_2Cu_3O_{7-x}$ ceramics of increased brittleness: *1* brittle fracture, *2* weak plasticity

(Fig.4.1, curve *1*). As the temperature is increased, the deformation behavior alters for the majority of the samples under investigation [4.9, 11, 24], and in the region of a particular temperature $T_0$ the transition to plastic flow is observed (Figs.4.1, 2, 5). For the majority of the investigated samples the brittle-plastic transition temperatures $T_0$ fall in the temperature region of the orthorhombic-tetragonal phase transformation [4.44].[1] As this temperature is approached, the deformation curve begins to exhibit, before fracture, a noticeable deviation from a quasielastic slope (Fig.4.1, curve *2*). At higher temperatures the plasticity of the samples increased drastically, and the deformation curves usually assumed a characteristic form with a "yield tooth" (Fig.4.1, curve *3*).

The transition to plastic flow is accompanied by a sharp decrease of the deforming stress and, in particular, of the stress $\sigma_c$ at which brittle fracture or a noticeable plastic flow occurs [4.9]. The temperature dependence of $\sigma_c$ for a series of Y-Ba-Cu-O ceramic samples are shown in Fig. 4.5. The $\sigma_c$ level depended on the synthesis conditions and the initial structure of the samples. In what follows (Sect.4.2.4), we shall demonstrate the influence of the oxygen amount on this parameter. All these factors, to some degree, affect the brittle-to-plastic transition temperature, too. Thus, the temperature $T_0$ depends not only on the chemical composition of HTSCs but also on the sample-preparation method.

The transition from brittle fracture to plastic flow occurs in a narrow temperature range near $T_0$. This behavior is typical for Y-Ba-Cu-O ceramics ($T \simeq 1000 \div 1100$ K), but this is not the only situation that can be observed when studying these and other HTSC compounds [4.9,11-13,23]. There are, for example, Y-Ba-Cu-O samples in which the brittle-to-plastic transition is, practically, absent, and at all the temperatures investigated they demonstrate brittle or quasibrittle fracture (Fig.4.6). We should also point out the appreciable scatter of the ultimate strengths in these samples, which conceals a $\sigma_c$ drop in the temperature range above $T_0$. Having analyzed the factors causing such a behavior of this group of samples, we concluded that their increased brittleness may be attributed to a higher degree of ceramic grain coalescence and also to the occurrence, during synthesis, of a large fraction of microcracks. These factors, seemingly, determine the mechanical properties of HTSC ceramics of different composition.

Standard techniques are usually employed to study the plasticity of HTSC ceramics at elevated temperatures. We shall discuss here the results by *von Stumberg* et al. [4.12] and refer the reader to [4.13,23]. Subjecting the Y-Ba-Cu-O ceramics to compressive deformation within $1140 \div 1250$ K has shown that the dependence of the deformation rate $\dot{\epsilon} = d\epsilon/dt$ on the stress $\sigma$ and the temperature T is described by

$$\dot{\epsilon} \simeq \beta \sigma^n e^{-Q/RT} , \tag{4.1}$$

where $n \simeq 1$. Q and the factor $\beta$ depend on the partial oxygen pressure $p(O_2)$ in the ambient atmosphere [4.12]: $\beta \propto [p(O_2)]^{1/2}$ and $Q \simeq 800$ or 600

---

[1] We have to note that in accordance with the results by Murase et al. [4.24] plastification of the Bi-Pb-Sr-Ca-Cu-O ceramics is observable in the range of approximately the same temperatures. Therefore the coincidence of $T_0$ with the orthorhombic-tetragonal transformation temperature in the Y-Ba-Cu-O ceramics may be incidental. Though, undoubtedly, "softening" of grain boundaries may take place in the phase-transformation region.

kJ/mol with $p(O_2) \simeq 100$ and 3 kPa, respectively. There exist other estimates of the activation energy and $\beta$ as a function of the oxygen pressure [4.13, 23]. Analysis of these data suggests the diffusional mechanism in high-temperature deformation of HTSCs.

The assumption that the character of the high-temperature deformation is diffusional [4.12, 13, 23] is consistent, in particular with the data on the change of the phase composition and grain size in deformed samples [4.9, 12]. Deformation is always accompanied by structure modifications. At brittle fracture and weak plastic flow, and upon heating the samples without deforming them, the main phase composition, however, persists. More radical restructurings occur under plastic flow (Fig.4.1, curve *3*). For example, using an X-ray microprobe and X-ray diffractometry one could register the phase decomposition of $YBa_2Cu_3O_{7-x}$ with the formation of mainly $Y_2BaCuO_5$, $BaCuO_2$ and CuO in the process of deformation. We note again the absence of such processes upon heating the samples without deforming them, and assume that precipitation of these phases in the form of inclusions may be related to chemical ("mechanochemical") processes in the grain-boundary region upon crystallite migration. These processes may affect the deformation behavior and the form of stress-strain curves.

Grain boundaries play a special role in the plasticity of various ceramic materials, in particular HTSC ceramics. It is the grain-boundary glide rather than deformation of the constituent crystallites that mainly determines the high-temperature deformation of Y-Ba-Cu-O and other ceramics. This, in particular, is supported by metallographic data [4.9b, 11]. We may assert that under normal deformation conditions of ceramic samples the deformation processes in crystallites are not clearly evident. There is, however, a possibility to activate the deformation of ceramic crystallites by suppressing the crack-formation processes and the related fracture of the samples. We shall refer, e.g., to the results of *Rabier* and *Denanot* [4.25] in which the room-temperature deformation of Y-Ba-Cu-O ceramics was carried out in special chambers, involving, together with a compressive load, a confining hydrostatic pressure.

The deformation mechanisms and subsequent factors limiting plasticity of HTSC ceramics are, in general, analogous. Some features of the deformation are related to different degrees of their chemical stability at elevated temperatures. One may, for instance, point out the change of plasticity in Y-Ba-Cu-O and other 1-2-3 HTSCs as their oxygen content is varied [4.9]. We shall come back to such data in Sects.4.2.4 and 4.3.3, and now, in concluding this subsection, we shall discuss another possibility for the deformation of HTSC ceramics.

Along with the deformation under uniform heating, Y-Ba-Cu-O ceramics were locally deformed [4.9] in the region of local resistive heating [4.45]. Like in uniform sample heating, noticeable plasticity in the local-domain region arises after heating above the brittle-to-plastic transition temperature. Employing this phenomenon enables one to carry out local thermomechanical treatments of ceramic samples, thus manipulating the structure and properties of HTSCs.

### 4.2.3 Specific Features of Deformation of Single Crystals

In contrast to ceramics, data on macroscopic deformation of single crystals, due to the afore-mentioned reasons, are practically not available. Therefore, we report here only briefly the first experiments on the deformation of Y-Ba-Cu-O single crystals by bending them perpendicular to the basal *ab*-plane [4.9b].

The $YBa_2Cu_3O_{7-x}$ single crystals are bent in air or in an oxygen flux at temperatures below 1200 K. "As-grown" and "oxygen-rich" samples were employed. The single crystals were pellets ($\simeq 0.1 \times 2 \times 3\,mm^3$) which predetermines their deforming technique. It was found that within this temperature range the deformation of all the tested single crystals was followed by crack formation. These processes were particularly noticeable on the extented side of the bent crystals.

Figure 4.7 shows a deformation curve of an "as-grown" Y-Ba-Cu-O single crystal at $T \simeq 1100$ K. The ultimate degree of deformation for this sample is $\epsilon \leq 5\%$. The deformation behavior for single crystals is different from the plasticity behavior of ceramics (Fig.4.1, curve *3*). At all temperatures the single crystals under bending had lower plasticity than the ceramic samples. This again emphasizes the special role of grain boundaries in the deformation of ceramics.

Bending of single crystals perpendicular to the basal *ab*-plane is not the best technique to plastically deform them. Moreover, high-$T_c$ and, in particular, Y-Ba-Cu-O single crystals have a pronounced anisotropy of the electrical, magnetic and optical properties. This holds for their mechanical

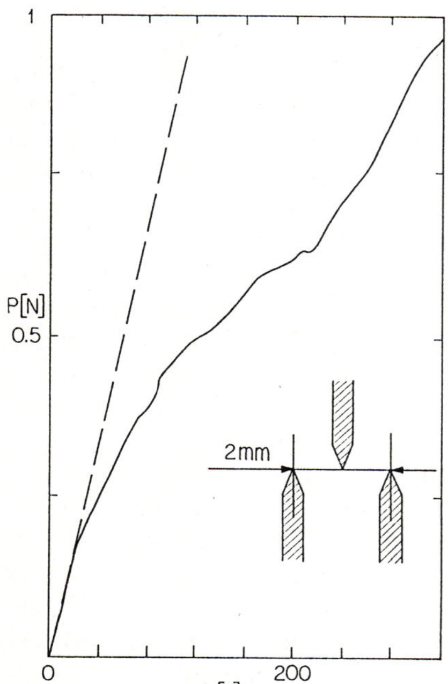


**Fig.4.7.** Deformation curve at three-point bending (shown at the *lower right*) of a $YBa_2Cu_3O_{7-x}$ single crystal perpendicular to the basal *ab*-plane. The bending was performed in air at $T \simeq 1100$ K at a rate of 10 $\mu$m/min. The *dashed line* shows the slope of elastic deformation

characteristics as well. For example, tests on the delamination of Y-Ba-Cu-O single crystals along the *ab*-planes indicate a weakness of the atomic bonds of the crystal lattice in this direction [4.46]. In accordance with existing ideas this may give rise to anisotropy of the plastic properties [4.14, 16, 17, 43]. A more detailed study of the deformation processes would require bigger single crystals, the synthesis of which is a goal for the near future. There are already publications from *Emel'chenko* et al. [4.47, 48], reporting on the synthesis of massive single crystals of $La_{2-x}Sr_xCuO_4$ and $La_2CuO_4$. Synthesis of $YBa_2Cu_3O_{7-x}$ single crystals measuring about 2x6x6 $mm^3$ and of even larger Y-Ba-Cu-O single crystals has also been reported [4.47-49]. This raises hopes for obtaining fresh data on the mechanical properties of HTSC single crystals and other perovskites, and also on the influence of defects on their various physical properties.

### 4.2.4 The Influence of Oxygen Content

The oxygen content in $YBa_2Cu_3O_{7-x}$ compounds and their analogues may be smoothly varied within $0 \le x \le 1$ by a thermal treatment. This gives a unique possibility to study the influence of the oxygen factor on the mechanical characteristics of the HTSC ceramics and single crystals and, in particular, to investigate the plastic properties of the materials in the phase-transformation region [4.50-52]. We have already discussed the dependence of the plastic-flow rate in Y-Ba-Cu-O ceramics at high temperatures on the partial oxygen pressure in the ambient atmosphere, see (4.1) in Sect. 4.2.2 [4.13, 23]. Now we shall consider the data on the influence of oxygen content on the quasielastic deformation before fracture and on the ultimate strength [4.9].

In the temperature range below the brittle-to-plastic transition of Y-Ba-Cu-O ceramics and other 1-2-3 compounds, a decrease of the oxygen concentration gives rise to their plastification. This can be inferred from the data in Fig.4.8. It is seen that a decrease of the oxygen concentration in $YBa_2Cu_3O_{7-x}$ (or, in other words, an increase of the oxygen deficit index x) leads to a drop in the ultimate strength $\sigma_c$ and the occurrence of a portion with a noticeable plasticity on the deformation curve before fracture. More detailed data on the influence of the oxygen amount on the ultimate strength in Y-Ba-Cu-O are shown in Fig.4.9. Note the non-monotonic behavior of $\sigma_c(x)$ with a minimum near $x \simeq 0.5 \div 0.6$. In this x range the 1-2-3 compounds are known to undergo the orthorhombic-tetragonal phase transformation [4.50]. We may suppose that $\sigma_c(x)$ reflects the influence of the oxygen content on the material state in the grain-boundary region.

There is a correlation between the influence of the oxygen amount on the deformation behavior and on the crystal lattice parameters of $YBa_2Cu_3O_{7-x}$. This is supported by the data of Fig.4.10. In the starting state the $YBa_2Cu_3O_{6.95}$ sample has the lattice parameters: $a_0 \simeq 3.83$, $b_0 \simeq 3.88$ and $c_0 \simeq 11.67$ Å. Figure 4.10 demonstrates the change of these parameters with increasing oxygen deficit ($x \simeq 0.5 \mp 0.1$ and $x \simeq 0.95 \mp 0.05$) and the analogous data after the deformation of these samples. The curves

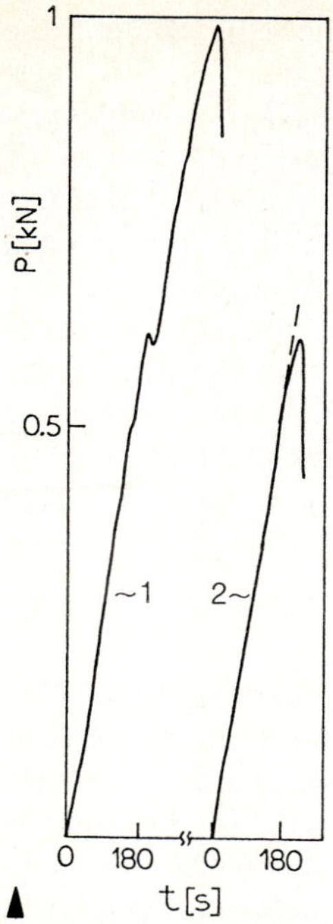


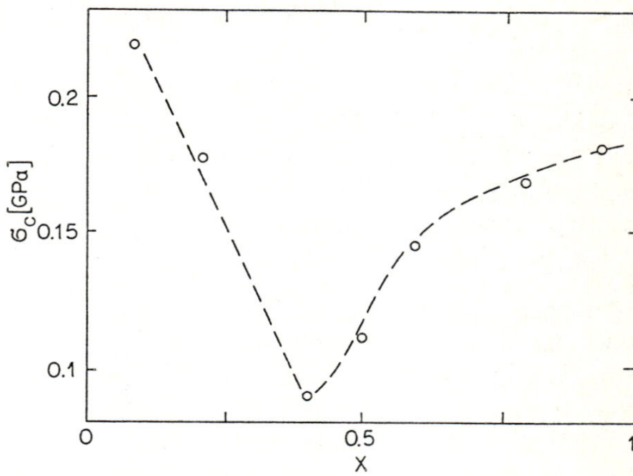


▲
**Fig.4.8.** Deformation curves of two oxygen-deficient $YBa_2Cu_3O_{7-x}$ ceramic samples, $x \simeq 0.05$ (*1*) and 0.5 (*2*). The compressive deformation was at a rate of 100 $\mu$m/min, $T \simeq 960$ K

**Fig.4.9.** Ultimate strength of $YBa_2Cu_3O_{7-x}$ ceramics as a function of oxygen deficit x. The samples were deformed in a helium atmosphere at $T \simeq 960$ K

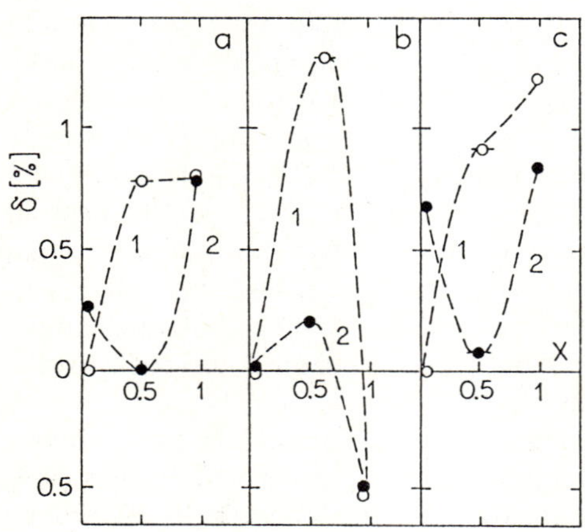


**Fig.4.10.** Relative change $\delta$ in crystallographic parameters *a* (**a**), *b* (**b**) and *c* (**c**) upon deformation of $YBa_2Cu_3O_{7-x}$ ceramics with different oxygen deficit x: *curves 1* - before deformation *open circles*, *curves 2* - after deformation at $T \simeq 960$ K (*filled circles*)

reveal the trends for the relative changes in *a*, *b* and *c* given by, e.g., for *a*, $\delta = (a-b_0)/a_0$. Note the decrease of the crystal-lattice parameters and, particularly, of the parameter *b* under the plastic deformation of sampleas with

$x \simeq 0.5$. We also note the approximate coincidence of the scale of $\delta$ and the degree of strain $\epsilon$ (about 1%).

With $x \simeq 1$, the deformation curves of the $YBa_2Cu_3O_{7-x}$ ceramics assume a form close to that of the starting samples (e.g., Fig.4.8, curve *1*), i.e., the samples become brittle again. The change in the crystal lattice parameters after the deformation in this case is also less pronounced (Fig.4.10). An analysis of these data suggests that plastification of the Y-Ba-Cu-O ceramic and other 1-2-3 compounds with intermediate x values is related to the oxygen reconstruction (its vacancies) upon deforming the sample. However, in order to draw a firmer conclusion about the role of oxygen in the processes of deformation, further studies need to be conducted. We shall come back again to this question during an analysis of the microhardness data in Sect. 4.3.3.

### 4.2.5 Structural Analysis

Structural investigations of deformed samples are important for the understanding of the nature and mechanisms of the HTSCs' deformation and for elucidating the influence of various defects of the crystal structure on different properties of high-$T_c$ superconductors. The activity in this field is not large, but the available data already suggest conclusions about the main structural modifications occurring in the process of deformation [4.5,9, 12,13,24-27,53-56].

Deformation of HTSC single crystals and ceramics gives rise to various structural defects. For example, the processes of crack formation and the occurrence of disturbances in the boundary region during grain-boundary glide are significant for the high-$T_c$ superconductor technology [4.57]. We have outlined in Sect.4.2.2 the change of the phase composition under high temperature deformation of Y-Ba-Cu-O ceramics [4.9]. Here, we shall add the influence of the deformation on the concentration and position of oxygen in the crystal lattice (Sect.4.2.4) [4.5,9]. The deformation of HTSCs also leads to a change in the twin structure [4.5,9,54,55], and the formation of dislocations and various point defects [4.9,12,13,24-27,53]. Naturally all the structural changes are of great concern, but the scope of this section does not allow us to give them equal consideration. We shall mainly discuss data associated with classic deformation carriers, i.e., dislocations [4.17,43]. We shall comment on the data concerned with the influence of mechanical actions on the twin structure, remembering that we shall come back to this matter in Sects.4.3.4 and 4.3.5 (see also Chap.6).

Several researchers have reported on Transmission Electron Microscopy (TEM) studies of the dislocation structure in deformed HTSCs [4.9, 24-27, 58]. We would also like to mention a data analysis by *Smirnova* [4.59). As an example, we present the results of TEM observations of a dislocation band in a crystallite of the $YBa_2Cu_3O_{6.95}$ ceramic after brittle fracture at $T \simeq 700$ K (Fig.4.11) [4.9]. It consists of dislocations with pronounced edge and screw components that intersect the system of twins without forming macroscopic jogs at their boundaries. This might indicate

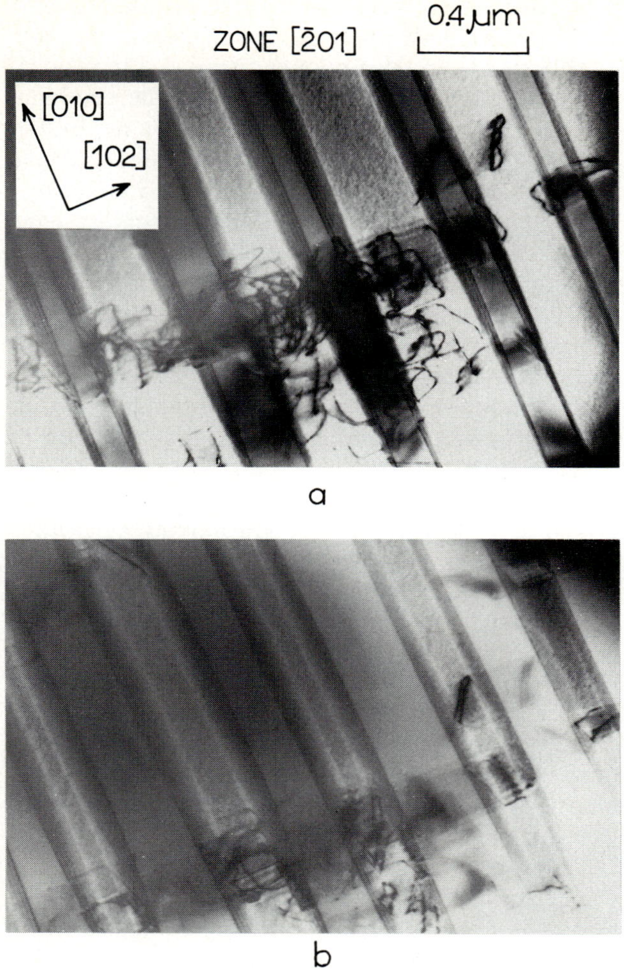


**Fig.4.11.** $YBa_2Cu_3O_{6.95}$ ceramic after brittle fracture at T $\simeq$ 700 K (transmission electron microscopy). (**a**) The image of the dislocation glide band, intersecting the twin system in one of the ceramic crystallites is seen in the [2̄01] zone. (**b**) The dislocation image is partly extinguished in the reflection (020). A peculiar contrast of a dislocation glide plane trace is observed.

that twin boundaries are poor obstacles for mobile dislocations. The results by *Rabier* and *Denanot* [4.25], on the contrary, may indicate a strong interaction between dislocations and twins.

The dislocation-band image in Fig.4.11 is partly extinguished in the (020) reflection (Fig.4.11b). An extinction method has been employed to estimate the Burgers-vector direction of dislocation in the glide band; it corresponds to the [010] direction. A dislocation with the ⟨100⟩-Burgers vectors in a Y-Ba-Cu-O crystal was already reported in [4.25-27, 58, 59]. We also mention that after high-temperature deformation of Y-Ba-Cu-O single crystals in the tetragonal phase by means of indentation *Yoshida* et al. [4.27], along with the ⟨100⟩ dislocations, observed dissociated dislocations with the Burgers vector ⟨110⟩ which, in their opinion, arise due to dislocation reactions. In this respect it should be pointed out that the results of *Muraset* et al. [4.24] show that ⟨110⟩ dislocations are typical for the Bi-Pb-Sr-Ca-Cu-O compounds. Vice versa, ⟨100⟩ dislocations are observable in this compound's crystallites only after annealing of the deformed samples.

Using the data of Fig.4.11, we may determine the dislocation-band glide plane for the Y-Ba-Cu-O samples investigated. To this end we shall use the fact that the leading dislocations in the band are sometimes stopped at twin boundaries, thus leaving a trace behind. This makes it possible to determine the angle of intersection of glide and twinning planes. These data and the estimation of the Burgers-vector direction enable us, with the help of geometrical calculations relevant to projections of twinning and glide planes onto the plane of observation, to unambiguously estimate the dislocation-band glide system to be ⟨100⟩ {100} which agrees, e.g., with the data of [4.25].

The TEM displayed in Fig.4.11b illustrates how a moving dislocation leaves behind an obvious glide-plane trace that has a peculiar contrast. A glide-plane trace is observed even in the absence of dislocations, that is, it is not associated, e.g., with a usual contrast from dislocation dipoles. One may suppose that this contrast is related to atomic displacement during the motion of dislocations. Taking into account a high oxygen mobility in Y-Ba-Cu-O and other 1-2-3 compounds, one may suppose that atoms of precisely this element are displaced during the dislocation motion. This supposition is qualitatively consistent with the results of the preceding section, which shows the dependence of the deformation processes on the oxygen amount [4.9]. We shall come back to these results in Sect.4.3.3 when discussing the influence of the amount and state of oxygen on the processes of deformation induced by indentation of the $YBa_2Cu_3O_{7-x}$ crystals [4.5, 53]. Now we would like to point out that there exist different factors which may give rise to point defects in the process of dislocation motion [4.17, 43]. In the case of the HTSC compounds specific processes may take place which are associated with their low stability with respect to oxygen, the complexity of the composition and configuration of dislocation cores [4.59].

The data in Fig.4.11 pertain to the temperature range below the brittle-to-plastic transition in Y-Ba-Cu-O ceramics. In this case the dislocation bands are apparently nucleated in the proximity of the stress concentrators (e.g., in grain-contact sites), their concentration in the crystallites is not large. Another situation arises at a high-temperature deformation of the HTSCs [4.9, 24, 27, 58]. We shall illustrate it by the TEM results in Fig.4.12 for a Y-Ba-Cu-O ceramic after plastic deformation at $T \simeq 1150$ K [4.9b]. In this case the ceramic exhibits the appearance of crystallites in which twins are absent and a dislocation network forms. No dislocation bands were observed in these crystallites. This situation is characteristic of the tetragonal Y-Ba-Cu-O single crystals [4.27]. We cannot say yet, however, whether the appearance of these crystallites is related to the orthorhombic-

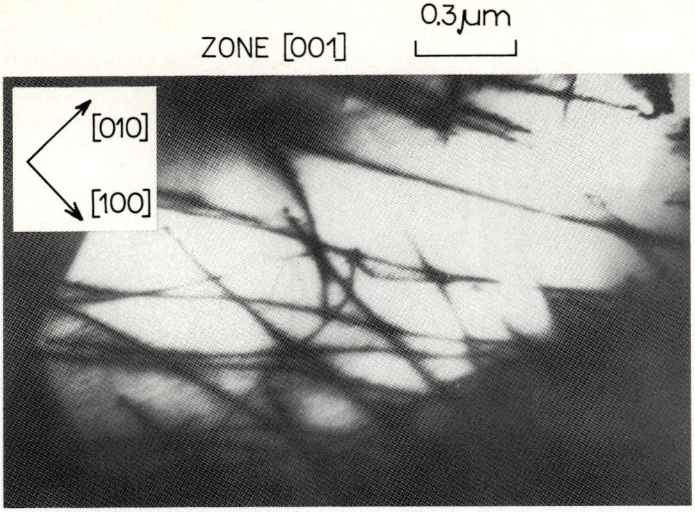


**Fig.4.12.** A dislocation network in an untwinned ceramic crystallite of $YBa_2Cu_3O_{0.95}$ after plastic deformation at T $\simeq$ 1150 K; TEM in the [001] zone

to-tetragonal phase transformation [4.44] or to the effect of mechanical untwinning [4.5,9,54,55]. But, undoubtedly, the processes of high-temperature deformation of HTSCs are followed by twin structure alterations and dislocation processes.

Twins are one of the main structural elements of high-$T_c$ superconductors, particularly of 1-2-3 compounds [4.5,25,54-56,58,59] (Chaps.1-3, 6). They may affect various physical properties of HTSCs, their mechanical characteristics included. Therefore, data on the influence of different actions on the twin structure are significant. We have already given an example of one such influence in Fig.4.12. Now we shall discuss some other data. First, we shall concentrate upon the effect of detwinning induced by indentation of Y-Ba-Cu-O single crystals observed by *Bobrov* et al. [4.5] and *Dorosinskii* et al. [4.54]. Quite the opposite effect was observed upon indentation of Y-Ba-Cu-O crystals, namely, a generation of twins in the vicinity of the indentation [4.5]. These phenomena are related to local changes in the twin structure, and we shall illustrate them in Sects.4.3.4 and 4.3.5. They are analogous to the well-known data obtained for ferroelectrics [4.60]. Attempts at achieving complete untwinning in HTSC crystals under uniaxial compression are also of great interest [4.5,55,56] (see also Chap.6). Alteration of the twin structure has also been observed under deformation of Y-Ba-Cu-O single crystals by bending (Sect.4.2.3).

Before concluding this extensive section and turning to deformation of HTSCs by means of microindentation technique, we shall mention the role of twins in deformation processes. It should be noted that disorientation of the lattice in neighboring twin domains is small (Chaps.3,6). The deformation of HTSC crystals under twinning or detwinning is small as well. One can confirm this, not only by calculations but also by direct measurements under indentation, bending, or uniaxial compression. The interest in twins is, to a great extent, associated with their influence on other deformation mechanisms [4.25] and, which seems to be more important for HTSCs, with their influence on the electric and magnetic properties (Sect.4.4 and Chaps. 5,6).

## 4.3 Microhardness

A specific feature of the indentation technique is that it enables one to measure local plastic characteristics and to avoid difficulties encountered when studying the deformation of HTSCs (Sect.4.1). Therefore, this technique is widely employed to study the mechanical properties of these materials. In a majority of such measurements, Vickers microhardness $H_v$ was used as a characteristic of the mechanical properties of the HTSC ceramics and single crystals [4.4-7,53,61-64]. Some researchers used indentation to investigate the processes of crack formation and to measure brittle-fracture parameters [4.4,6,64].

In this section we shall focus on the data on microhardness of the HTSC ceramics and single crystals, specifically of the Y-Ba-Cu-O. A considerably smaller number of reports on the use of the microindentation technique in studies of brittle fracture processes are known. We shall restrict ourselves to just mentioning some of the observations. Before discussing the results, we shall dwell on some questions associated with the technique.

In microhardness studies the surface of the materials is indented by means of a diamond pyramid. Then the indentation sizes are measured and the data are statistically processed. The Vickers microhardness $H_v$ is determined from the relationship [4.65,66]:

$$H_v = 1.854\,P(2d)^{-2} \,, \tag{4.2}$$

where P is the loading on the indenter, 2d is the indentation diagonal length (Fig.4.13). To correctly employ (4.2) one has to verify its applicability. An example of such a verification for a Y-Ba-Cu-O single crystal is given in Fig.4.14 [4.6a]. Equation (4.2) is seen to hold in the range $0.05 \leq P \leq 0.5$ N. With lower loading $H_v$ fictitiously increases, which is related to the elastic recovery of the indentations [4.65-67]. With high loading, in contrast, one may encounter a decrease of $H_v$ due to crack formation. In Y-Ba-Cu-O single crystals in the range of $P \leq 1$ N cracks preferentially form at the indentation angles, and the dependence of $H_v(P)$ is weak. Thus, (4.2) applied to single crystals is valid in the range $P \simeq 0.05$ to 1 N. For ceramics the range of P depends on the synthesis conditions and the crystallite dimensions. For standard samples with the crystallite dimension $D \sim 10\ \mu$m the grain-boundary destruction begins with $P \geq 0.3 \div 0.5$ N. Therefore the data, which will be discussed below, were obtained mainly with indenter loadings in the range $P = 0.1 \div 0.35$ N. Microhardness measurements of ceramic sam-

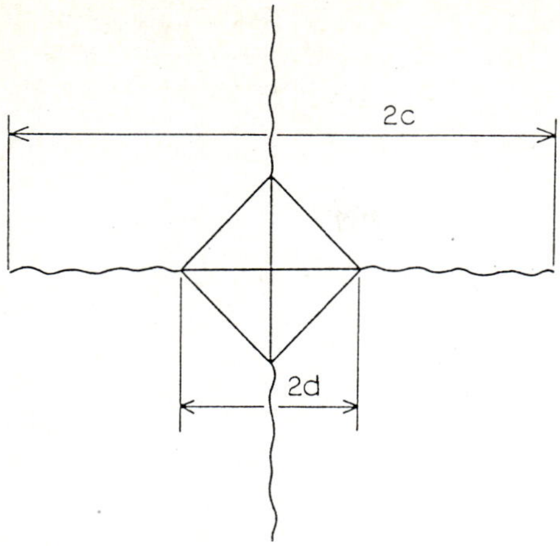


**Fig.4.13.** Indentation (schematized) and the parameters to be measured in microhardness and crack formation tests

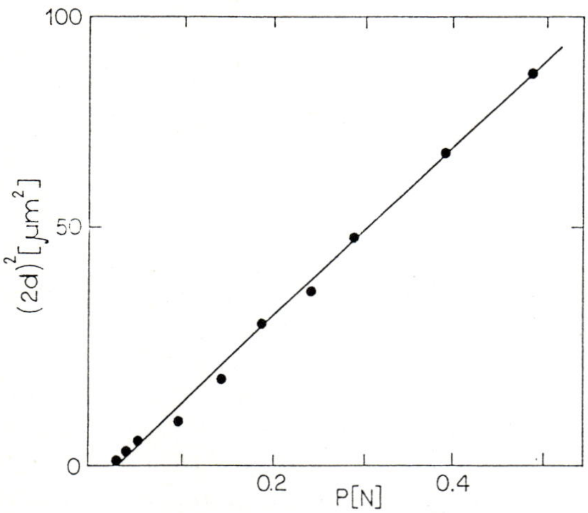


**Fig.4.14.** Length of the indentation diagonal on the surface of a $YBa_2Cu_3O_{7-x}$ single crystal as a function of the indenter loading P. The indentation diagnonal is parallel to one of the ⟨110⟩ directions

ples at higher loading are not quite correct and may serve only as a qualitative indicator to the strength properties of ceramics.

### 4.3.1 Ceramics and Single Crystals. Data Comparison

The microindentation technique enables one to judge the homogeneity of the mechanical properties of materials. For example, the scatter in microhardness values may be affected by heterogeneity of the phase composition. This situation was observed by *Lubenets* et al. [4.63] in Bi-Sr-Ca-Cu-O single crystals. In contrast to single crystals, in ceramics (even with homogeneous phase composition) the $H_v$ value may be affected by grain boundaries [4.53].

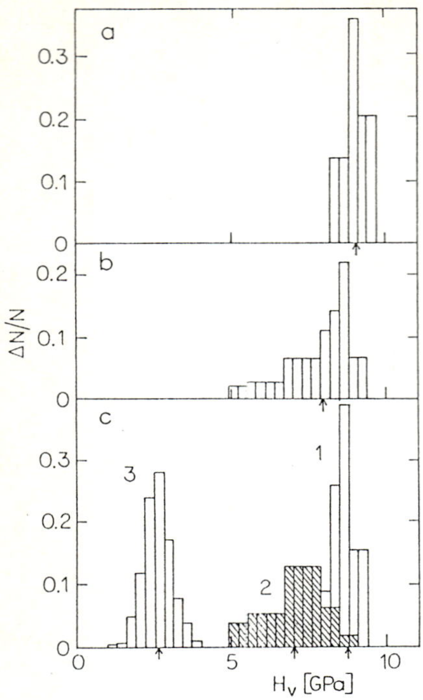


**Fig.4.15.** Microhardness histograms of $YBa_2Cu_3O_{6.95}$ single crystals and ceramics: (**a**) single crystal; (**b**) coarse-grained ceramics under continuous indentation; (**c**) data for crystallites (*1*) and grain boundaries (*2*) of coarse-grained ceramics of part (**b**) and data for fine-grained ceramics (*3*). The samples are indented at $T \simeq 300$ K, $P = 0.15$ N (**a** and **b**) and $P = 2$ N (**c**, histogram *3*). The *arrows* indicate mean microhardness values

In order to discuss the effect of grain boundaries on the microplastic chracteristics we shall compare the data for $YBa_2Cu_3O_{6.95}$ ceramics and single crystals, homogeneous in phase composition. We shall employ the results by *Fomenko* et al. [4.53] presented in the form of histograms of $H_v$ as in Fig.4.15. The Y-Ba-Cu-O ceramics and single crystals were indented at room temperature, and the histograms were plotted on the basis of 50÷500 indentations for each sample. For single crystals the deviation of the length of the indentation diagonal from a mean value did not exceed the optical system error, i.e., the histogram of the $H_v$ values distribution (Fig.4.15a) reflects only the accuracy of the measurements. The histogram in Fig.4.15b was plotted from measurements of indentations applied onto the ceramic surface with continuous indentation and the same indenter loading, as in the case of the single crystal ($P = 0.15$N). Note the asymmetry of this histogram with broadening towards low $H_v$ values. This is precisely what reflects the influence of grain boundaries. In fact, if we separate the obtained data into two groups, one involving indentations inside grains, the other indentations in the vicinity of grain boundaries, then the corresponding histogram will show up as two symmetric parts (Fig.4.15c, curves *1* and *2*) the former being close to the histogram for single crystals (Fig.4.15a). So, when the $H_v$ values are measured in the central region of ceramic crystallites ($d \ll D$), they may be almost comparable to those for single crystals. Mean $H_v$ values of crystallites are somewhat lower than those for single crystals. Here the influence of the near-boundary regions may be manifested. Deviations of the crystallite indentation planes from the basal orientation may be important. The same factors, apparently, lead to a more noticeable data scatter for ceramics than for single crystals.

Histogram *2* in Fig.4.15c shows lower values and a noticeable dispersion of the microhardness in the boundary region of crystallites. The influence of grain boundaries is illustrated also by the data for fine-grained ceramics with P = 2 N (Fig.4.15c, histogram *3*). With such loading, the indentations embrace the area of several grains, and active crack formation is observed. The microhardness in this case may be measured only relatively, the obtained $H_v$ estimates are nearly 3 times lower than those for ceramics at P = 0.15 N and, naturally, much lower than the $H_v$ of single crystals. These experiments suggest a lower $H_v$ and strength of the grain boundary region as compared with crystallites, which agrees with data obtained under the macroscopic deformation of HTSCs (Sect.4.2.1). The influence of grain boundaries determines the inhomogeneity of the mechanical properties of Y-Ba-Cu-O ceramics and other HTSC compounds.

### 4.3.2 Temperature Dependence of Microhardness

The temperature dependence of the mechanical properties of materials plays a significant role in studies of the processes which determine their plasticity. However, most of the data known so far on HTSC microhardness have been obtained at room temperature. Only several works are known [4.5,27,53,64] in which the indentation temperature was varied. The most detailed investigations were carried out by *Fomenko* et al. [4.53] with ceramics and Y-Ba-Cu-O single crystals in the temperature range below room temperature. Attempts have been made to study $H_v(T)$ for single crystals at elevated temperatures [4.5]. This subsection is mainly devoted to analyzing the data of [4.5,53].

First we shall consider the results obtained in the range of $T \simeq 77 \div 300$ K [4.53]. For these temperatures the physical properties of $YBa_2Cu_3O_{7-x}$ exhibited some singularities. In Sect.4.2.1 we have already mentioned the anomaly of sound absorption [4.28-35] and microcreep [4.8,19-21]. Also, one should consider the reports on antiferromagnetic transformations in this range [4.68], the characteristic features of thermal properties [4.69] and so on. Finally, it should be noted that the critical temperature values for ceramics and $YBa_2Cu_3O_{6.95}$ single crystals lie within this temperature range, too [4.53].

Figure 4.16 exhibits the dependence of $H_v$ on T for crystallites and for the grain-boundary region averaged over several series of grains on a single sample of a coarse-grained $YBa_2Cu_3O_{6.95}$ ceramic. In both cases the microhardness can be seen to increase monotonically within the data spread with decreasing temperature ($\Delta H_v/H_v \simeq 30\%$); here the increase in $H_v$ exceeds the variation in the Y-Ba-Cu-O elastic moduli [4.26]. These data also provide evidence that at all the temperatures the maximum values of $H_v$ for ceramic crystallites are close to those of single-crystal microhardness, whereas the values of $H_v$ for grain-boundary regions are always below the corresponding values for crystallites.

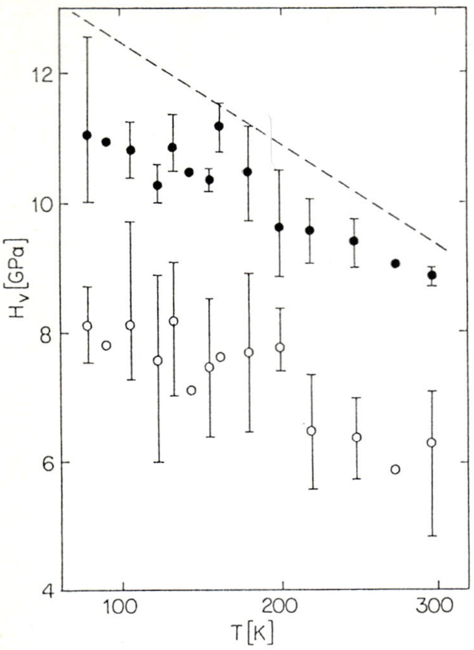


**Fig.4.16.** Temperature dependence of the microhardness $H_V$ for crystallites (*filled circles*) and intergrain boundary region *open circles* of coarse-grained ceramics ($YBa_2Cu_3O_{6.95}$) P = 0.15 N. The results are averaged with respect to several measuring cycles on one sample. The data scatter between minimum and maximum $H_V$ values is indicated by the *error bars*. *Dashed line*: averaged data for several single crystals for the same P

According to the differential magnetic-susceptibility data, ceramic samples have narrow superconducting transitions ($T_c \simeq 93$ K, $\Delta T_c \simeq 2 \div 5$ K). The results for the temperature range $T \leq 90$ K (Fig.4.16) thus refer to a superconducting state. Therefore, unlike microcreep [4.8a, 21, 22] (Sect. 4.2.1), in these experiments, within the data spread, no effect of the electronic state of high-$T_c$ superconductors on the processes of their deformation has been observed.

To some extent this confirms the conclusion, drawn in Sect.4.2.1, that such an effect under microcreep is associated with some peculiarities of the Y-Ba-Cu-O ceramic state in the grain-boundary region [4.21]. Note that $H_V(T)$ has no marked features in the temperature range within which the above anomalies of sound absorption, microcreep and other features of the HTSC properties have been observed.

The function $H_V(T)$ for $YBa_2Cu_3O_{6.95}$ and $La_2CuO_4$ single crystals does not exhibit any singularities either (Fig.4.17) [4.53].[2] For the whole temperature range analyzed, this dependence is linear, which may prove the thermoactivated behavior of deformation of the materials studied. Indeed, as has been shown by *Boyarskaya* et al. [4.66], the temperature dependence

[2] As shown by recent studies [4.53b], the $H_V(T)$ dependence of the La-Sr-Cu-O system behaves analogously. It should also be noted that other HTSCs reveal smaller values for $H_V$ (higher plasticity) of single crystals for one of the phases of the Tl-Ca-Ba-Cu-O compounds (see also the data of [4.63] for Bi-Sr-Ca-Cu-O crystals).

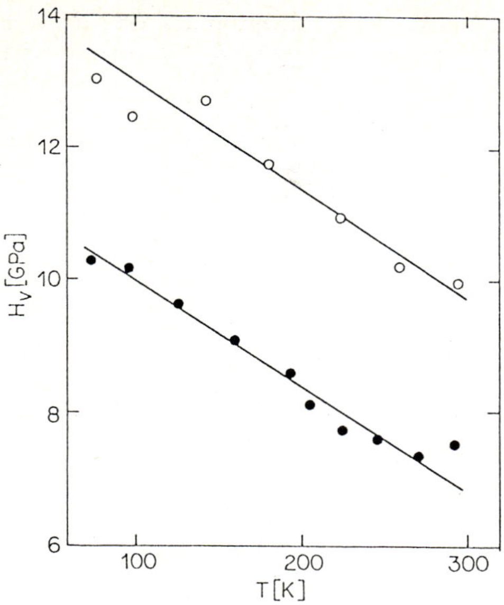


**Fig.4.17.** Temperature dependence of microhardness of $YBa_2Cu_3O_{6.95}$ (*open circles*) and $La_2CuO_4$ (*filled circles*); P = 0.15 N; both in the form of single crystals

of the indenter pattern variation rate for plastic materials is well described by the Arrhenius equation:

$$\dot{\epsilon} = \dot{\epsilon}_0 \exp[-U(\sigma)/kT] . \qquad (4.3)$$

Then, by using the approximation of a constant activation volume $U \simeq U_0 - \gamma\sigma$ for the activation energy, we may write out the relation for the microhardness temperature dependence [4.66, 67]:

$$H_v = \beta[(U_0/\gamma) - (kT/\gamma)\ln m] , \qquad (4.4)$$

Here $U_0$ is the activation energy under the stress $\sigma$=0, $\gamma$ is the activation volume, $\beta$ is the factor involving the Schmidt factor [4.17], the coefficient of proportionality between $H_v$ and yield stress $\sigma_c$ [4.66, 67] and the relation $m = \dot{\epsilon}_0/\dot{\epsilon}$. The function (4.4) agrees well with our experimental data.

Using (4.4) one may estimate the activation parameters $U_0$ and $\gamma$ for deformation processes in $YBa_2Cu_3O_{6.95}$ and $La_2CuO_4$ single crystals (Table 4.1). The values of $U_0$ and $\gamma$ depend on the choice of the parameters $\beta$ and m, the principal ambiguity is related to the parameter m. The estimates presented in Table 4.1 are made using $\beta \simeq 6$ and $\ln m \simeq 20$ [4.67] and the Burgers vector b = 3.82 Å for Y-Ba-Cu-O, based on [4.25]. Table 4.1 also presents extrapolated values of $H_v$ for Ge [4.67]. One can see that the microhardness and activation parameters of Y-Ba-Cu-O and La-Cu-O single crystals are close to those typical of semiconductor materials.

In discussing the plastic properties of materials at moderate temperatures, deformation are usually viewed to occur via dislocation mechanisms [4.14, 16, 17, 66, 67]. The estimates of the activation parameters given above

**Table 4.1.** Microhardness $H_v$ and estimates of the activation energy $U_0$ and activation volume $\gamma$ for $YBa_2Cu_3O_{6.95}$, $La_2CuO_4$ and Ge single crystals (*b* is the Burgers vector and $b^3$ its third power)

| | $YBa_2Cu_3O_{6.95}$ | $La_2CuO_4$ | Ge [4.67] |
|---|---|---|---|
| $H_v(T \simeq 300K)$ [GPa] | 10.0 | 7.5 | 7.5 |
| $H_v(T \simeq 0K)^a$ [GPa] | 14.6 | 11.8 | 19.0 |
| $U_0$ [eV] | 1.6 | 1.3 | 1.5 |
| $\gamma$ [$10^{-24}cm^3$] | 103.5 | 103.5 | 80.0 |
| $\gamma$ [$b^3$] | 1.8 | | 1.2 |

a Extrapolated data

for Y-Ba-Cu-O and La-Cu-O single crystals do not contradict this representation. Direct observation of dislocations in Y-Ba-Cu-O single crystals after high-temperature indentation has been achieved by the TEM technique developed by *Yoshida* et al. [4.40]. The preliminary TEM data by *Fomenko* et al. [4.53] show that the dislocation bands near scratches and indenter patterns also arise after indenting Y-Ba-Cu-O single crystals at room temperature. The observations of dislocations and glide bands in HTSC-ceramic crystallites (Sect.4.2.5) [4.9b, 24-26] are also consistent with these data. All the above-mentioned data may considered from the viewpoint of generation and thermoactivated motion of dislocations.

We have discussed the microhardness data in the range below room temperature, which are most complete. As mentioned in Sect.4.1, for elevated temperatures problems exist connected with both the indentation technique and the instability of HTSCs with respect to the phase composition and the incorporated oxygen [4.9, 15]. In addition to the above mentioned TEM studies made after high-temperature indentation [4.27], we may, for

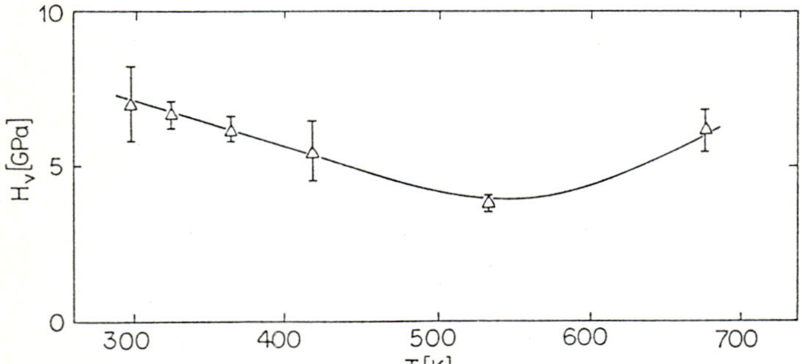


**Fig.4.18.** Microhardness $H_v$ as a function of the indentation temperature As-grown $YBa_2Cu_3O_{7-x}$ single crystals were indented in air; the measurements were conducted with P = 0.1 N

example, report the results on the dependence of $H_v$ on T, for "as grown" $YBa_2Cu_3O_{7-x}$ single crystals measured in air in the temperature range T $\simeq$ 300÷700 K (Fig.4.18) [4.5]. One can see that this dependence is nonmonotonic. The initial decrease in $H_v$ with increasing temperature is in qualitative agreement with the data considered in this section for the low-temperature region (Figs.4.16, 17). Further increase in $H_v$ in the T $\geq$ 500÷550 K range is no longer consistent with a thermoactivated behaviour of the deformation. In the following subsection we demonstrate that such a behavior of $H_v(T)$ is associated with a change in the state of oxygen upon heating the as-grown single crystals.

### 4.3.3 Effect of the Oxygen State

In this section we continue by first discussing the *high-temperature* indentation of as-grown $YBa_2Cu_3O_{7-x}$ single crystals [4.5] and then turn to the investigation of microplasticity of Y-Ba-Cu-O compounds with varying oxygen content.

To elucidate the participation of oxygen in processes determining the plasticity of HTSC materials we show the dependence of microplasticity of several as grown Y-Ba-Cu-O single crystals on the temperature $T_a$ of the preliminary annealing in oxygen (Fig.4.19) [4.5]. The microhardness of annealed samples was measured at room temperature. The curve $H_v(T_a)$ is seen to have a distinct minimum at $T_a \simeq$ 540 K. This makes it possible to assume that the analogous minimum on the temperature dependence of $H_v$ (Fig.4.18) is related to two factors: thermoactivated processes of deformation and alterations in the state of oxygen when as-grown crystals are heated.

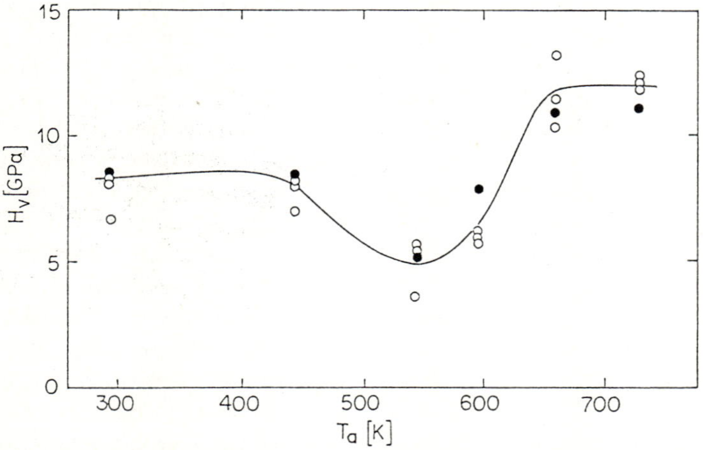


**Fig.4.19.** Microhardness $H_v$ of as-grown $YBa_2Cu_3C_{7-x}$ single crystals as a function of the annealing temperature $T_a$ in an oxygen atmosphere. Room temperature data of different samples (*filled circles*: special sample whose temperature dependences of the electric resistivity after annealing at different $T_a$ are shown in Fig.4.20)

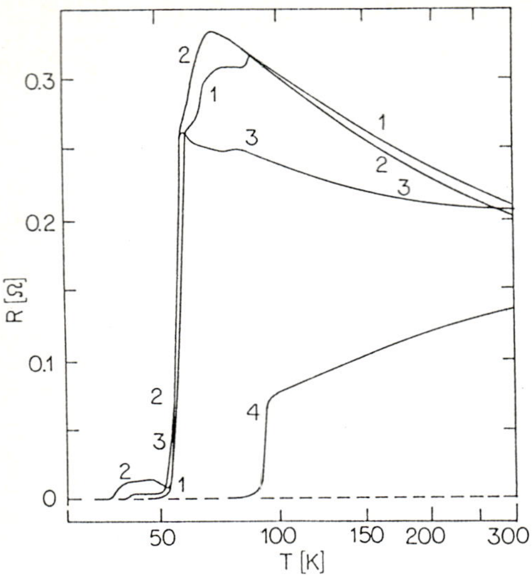


**Fig.4.20.** Temperature dependences of the electrical resistance of an as-grown $YBa_2Cu_3O_{7-x}$ single crystal after annealing in an oxygen atmosphere at different temperatures: *1*- $T_a$ = 140, *2*- 540, *3*- 650, *4*- 720 K (*filled circles* in Fig.4.19)

Some additional information on the oxygen-annealing effect on the state of as-grown single crystals can be obtained from simultaneous studies of plastic and electrical properties [4.5]. Figures 4.19 (*filled circles*) and 4.20 depict such a correlation. In the initial state the superconducting properties of this sample were mainly determined by the so-called 60° -phase ($T_c \simeq 60$K), whereas the 90° -phase ($T_c = 90$K) has been exhibited by the temperature dependence of the electrical resistance in the form of steps (Fig.4.20, curve *1*). Annealing in an oxygen atmosphere at $T_a \leq 440 \div 470$ K only slightly affected the electrical and mechanical properties but after annealing at higher temperatures, redistribution of these phases and variation in the properties mentioned were observed. A minimum of microhardness was noted after annealing at $T_a \simeq 540$ K, when the 90° -phase vanished as a result of oxygen redistribution, and the R(T) dependence exhibited a singularity arising due to formation of a new phase with $T_c \simeq 30 \div 40$ K (Fig. 4.20, curve *2*). All these phases are known to differ in their oxygen content [4.51, 70, 71]. In particular, the 30° -phase, according to the data of *Cava* et al. [4.51], has a high oxygen deficit. Further increase in the annealing temperature is accompanied by oxygen redistribution inside as-grown single crystals and its uptake from the ambient atmosphere. Saturation of samples with oxygen results in the vanishing of the 30° -phase and the appearance of the 90° -phase, and also in the conductivity "metallization" (e.g., Fig.4.20, curves *3* and *4*). It was against the background of these processes that further variation in $H_v$ took place: a sharp increase of $H_v$ was observed in the range of $T_a \simeq 570 \div 670$ K, and then at higher annealing temperatures, when the processes of oxygen saturation were completed, $H_v(T)$ approached saturation (Fig.4.19).

In Sect.4.2.4 it has already been noted that $YBa_2Cu_3O_{7-x}$ compounds and their analogs allow one to smoothly vary the oxygen concentration in the range of $0 \leq x \leq 1$. This enables one to investigate the effect of oxygen

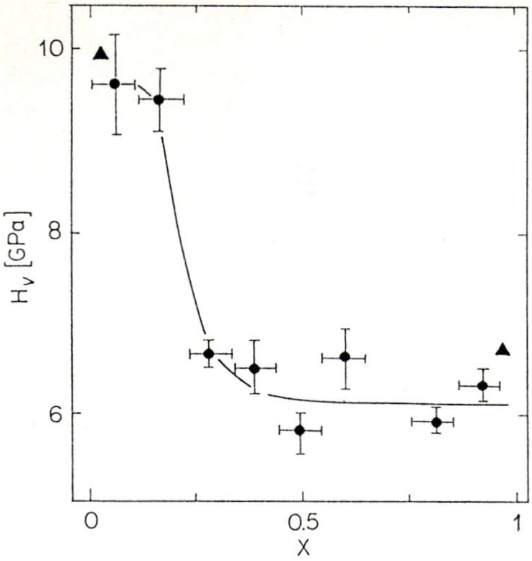


**Fig.4.21.** Microhardness $H_v$ of $YBa_2Cu_3O_{7-x}$ ceramic (*circles*) and single crystal (*triangles*) crystallites as a function of the oxygen deficit. The measurements were conducted at T $\simeq$ 300 K and P = 0.1 N. The estimation errors in the oxygen deficit x and mean-square deviations in $H_v$ estimates are indicated

content on deformation processes of these compounds and, in particular, to analyze mechanical characteristics at a fixed temperature in the region of the orthorhombic-tetragonal phase transition (x $\simeq$ 0.5) [4.50] and the so-called ortho-1-to-ortho-2 phase transition connected with vacancy ordering in the oxygen sublattice (x $\simeq$ 0.3÷0.4) [4.51,52]. To carry out such experiments *Fomenko* et al. [4.53] applied a special technique of proportioned variation in oxygen with a thermal treatment of ceramics and $YBa_2Cu_3O_{6.95}$ single crystals in an inert atmosphere and in vacuum.

The dependence of $H_v$ on the oxygen deficit x in ceramic crystallites is depicted in Fig.4.21. Here we also present the data for Y-Ba-Cu-O single crystals obtained at the limiting values of x $\simeq$ 0 and 1. Note, for instance, the drastic decrease (almost by a factor of 1.5) in $H_v$ in the range of x $\simeq$ 0.2÷0.4. The dependence of $H_v$ on x vanishes almost completely at other x values (x $\simeq$ 0÷0.2 and x $\simeq$ 0.4÷1) of oxygen concentrations. The data displayed are in qualitative agreement with the results of *Graft* et al. [4.72]. The quantitative difference of our data is likely to be associated with the fact that *Graft* used a larger loading during indenting and a different technique of sample preparation.

These data provide evidence that an abrupt change in $H_v$ occurs in that range of oxygen concentrations at which its atomic ordering changes in the Cu-O chains [4.51,52]. In contrast, no significant change in $H_v$ is observed at the concentration corresponding to a transition of the crystal lattice from orthorhombic to tetragonal [4.50]. The dependence of $H_v$ on x correlates with the data on internal friction observed by *Lemmens* et al. [4.36], who reported the change of the relaxation peaks at approximately the same values of x $\approx$ 0.3 [4.37-40]. This phenomenon is assumed to be related to changes in the conditions of atomic motion of oxygen in Cu-O chains. This transposition of oxygen atoms may proceed upon dislocation motion (e.g., in the core of mobile dislocations), and a change in the con-

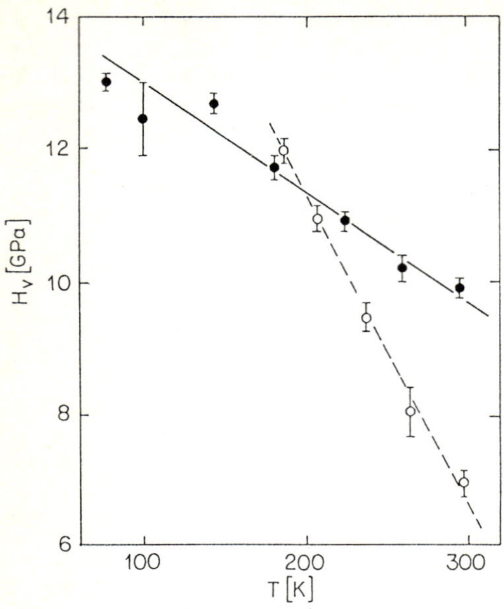


**Fig.4.22.** Temperature dependence of $H_v$ of $YBa_2Cu_3O_{7-x}$ single crystals containing different amounts of oxygen: oxygen-enriched single crystal, $x \simeq 0.05$ (*filled circles*); oxygen-deficient single crystal, $x \simeq 1$ (*open circles*); $P = 0.15$-$0.35$ N

ditions of their rearrangement may affect the dislocation mobility and deformation processes.

Alteration of the oxygen content affects both the microhardness value and its temperature dependence. This is illustrated by the data obtained for two single crystals with oxygen deficits $x \simeq 0$ and 1 (Fig.4.22) [4.53]. Both $H_v(T)$ curves in Fig.4.22 are linear in the range of $T \simeq 200 \div 300$ K, with no singularities observed. Removal of oxygen from Y-Ba-Cu-O crystals is accompanied by an increase in their plasticity at room temperature and by an enhancement of the temperature dependence of $H_v$. Measurements at oxygen deficit $x \simeq 1$ were made for the $200 \div 300$ K range only. At lower temperatures the single crystal exhibits an abrupt embrittlement; indenter patterns on its surface became smeared and cracks developed intensively. It was almost impossible to measure the microhardness of this sample at $T < 200$ K. The measurements at these temperatures were incorrect. For comparison, note that single crystals saturated with oxygen ($x \simeq 0$) retain rather high plasticity up to 77 K. Faceting of indenter patterns on their surfaces remains distinct at all the temperatures and radial cracks develop only in the corners of patterns.

Using the data of Fig.4.22 and (4.4) one may estimate, as has been done in Sect.4.3.2, the activation parameters and the influence upon them of the oxygen content in Y-Ba-Cu-O (Table 4.2). The estimates obtained show that removal of oxygen from Y-Ba-Cu-O single crystals results in a double decrease of the activation energy and in a triple decrease of the activation volume. A decrease in the activation energy facilitates deformation processes and leads to plastification of these compounds. At the same time, a decrease in the activation volume enhances the microhardness temperature dependence and embrittles single crystals at low temperature.

**Table 4.2.** Microhardness $H_v$ at $T \simeq 300$ K, the slope $\delta H_v/\delta T$ of $H_v(T)$ and estimates of the activation parameters $U_0$ and $\gamma$ for $YBa_2Cu_3O_{7-x}$ single crystals with oxygen deficits $x \simeq 0$ and 1

| | $x \simeq 0$ | $x \simeq 1$ |
|---|---|---|
| $H_v(T \simeq 300K)$ [GPa] | 10.0 | 7.0 |
| $\delta H_v/\delta T$ [$10^{-2}$ GPa/K] | 1.6 | 4.6 |
| $U_0$ [eV] | 1.6 | 0.8 |
| $\gamma$ [$b^3$] | 1.8 | 0.6 |

What is the cause of such an influence on deformation processes? We have already assumed that this phenomenon may be associated with the rearrangement of oxygen atoms during the motion of dislocations. If such is the case, the shift of dislocations may be followed by displacement of "oxygen" sites of the crystal lattice. Recall that the TEM exhibited a peculiar contrast in Y-Ba-Cu-O ceramic crystallites in the zone of motion of dislocation-glide bands (Sect.4.2.5., Fig.4.11). This may serve as an indirect support of the assumption of the displacement of oxygen atoms upon motion of dislocations, and diffusive processes may be essential here. At this stage, this question is unlikely to be answered more definitively. Further investigations are required to get an unambiguous solution of the problem.

To complete this subsection we examine once more the effect of oxygen on the anisotropy of the plastic properties of Y-Ba-Cu-O single crystals [4.5]. This anisotropy manifests itself in the dependence of microhardness on the orientation of the indenter-pattern diagonals relative to the crystallographic ones. For instance, the minimum values of $H_v^{\langle 100 \rangle}$ were recorded upon aligning the diagonals along the $\langle 110 \rangle$ direction (twin boundary orientations), and the maximum $H_v^{\langle 100 \rangle}$ were obtained upon their orientation along $\langle 100 \rangle$. The ratio $\Delta H_v/H_v = (H_v^{\langle 100 \rangle} - H_v^{\langle 110 \rangle})/H_v^{\langle 110 \rangle}$ may serve to characterize the anisotropy of $H_v$. This ratio changes when the oxygen content is varied and its state in the crystal lattice changed. As an example, we consider the data for as-grown Y-Ba-Cu-O crystals [4.5]. At the initial stage their microhardness anisotropy factor was usually $\Delta H_v/H_v \simeq 15 \div 20\%$ and, after saturation with oxygen at $T_a \simeq 670$ K, $\Delta H_v/H_v$ became as high as 50-80%. The ordering of oxygen in CuO1-O1 chains may be a cause of the anisotropy in the mechanical properties [4.70, 71]. It is worth mentioning that the effect of annealing on the anisotropy of $H_v$ was still observed in that range of $H_v$ where the dependence $H_v(T_a)$ was going to saturation (Fig.4.19). Therefore, after accomplishing the "metallization" of as-grown single crystals (e.g., Fig.4.20, curve *4*), when the concentration of the 90°-phase reached the level ensuring percolation, the processes of homogenization and oxygen ordering in the Y-Ba-Cu-O crystal lattice still went on, which is displayed by the enhancement of the microhardness anisotropy.

And lastly, in connection with the problem of the oxygen effect on the mechanical properties of HTSCs, one should bear in mind the qualitative difference in the data on the effect of oxygen deficit upon microhardness of crystallites and the critical breaking stress of Y-Ba-Cu-O ceramics (Figs. 4.9,21) [4.5,9,53]. This shows that the effects due to the oxygen factor on the strength and plasticity of crystals and on grain boundaries differ.

### 4.3.4 The Untwinning Effect

In Sect.4.2.5 we have already mentioned that different mechanical actions can affect the twin structure of Y-Ba-Cu-O and other HTSC compounds [4.5,9b,54-56]. Now we consider in more detail the results of observing the untwinning effect under indentation of Y-Ba-Cu-O single crystals [4.5,54] (see also Chap.6).

Figure 4.23 depicts optical polarizing photographs of two indented Y-Ba-Cu-O single crystals [4.5] with different densities of twin domains at the initial stage. The effect of untwinning ("monodomainization" of the struc-

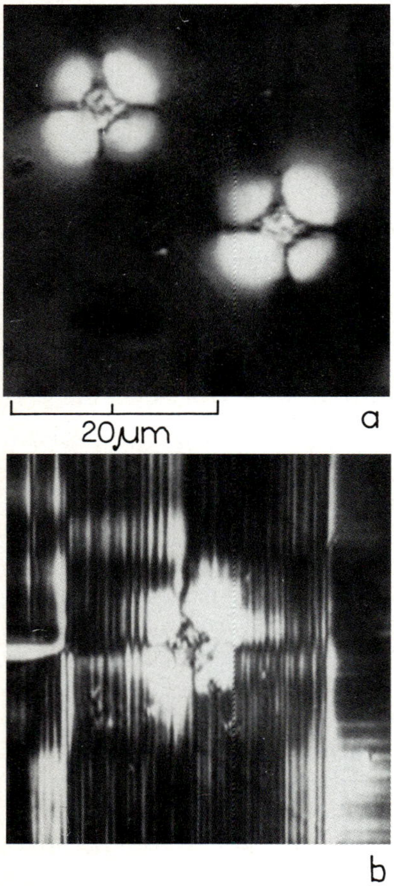


**Fig.4.23.** Polarization photographs of indented $YBa_2Cu_3O_{7-x}$ single crystals (**a** and **b** show samples with different initial twin densities). The light contrast near the indentor patterns is the region of untwinning

ture) in these photographs is exhibited in the form of light contrast near the indentation. The crystals with high density of twins, optically unresolved, were dark in the polarized light (Fig.4.23a). Upon rotating the sample in crossed polaroids, single-domain areas of crystals changed their color, whereas the coloring of the areas with high twin densities remained the same [4.73].

These are typical photographs illustrating the effect of untwinning in Y-Ba-Cu-O single crystals. Several years ago a similar phenomenon was observed by *Chernysheva* [4.60, 74] in ferroelectrics. We are prone to think that the nature of the untwinning mechanism in these materials is almost the same. ("Everything new is quite forgotten old".). Both classes of materials are polysynthetic twins [4.74], that is, their structure is a polydomain twin structure. Application of an external stress to crystals results in a displacement of twin boundaries and in enlargening twin domains of "favorable" crystallographic orientation [4.73, 74]. This is what causes untwinning in the stress field in the vicinity of the indentation. Such a rearrangement of the domain structure of an initially polysynthetic twin is presented schematically in Fig.4.24.

The kinetics of untwinning depends on the initial crystal structure. Samples with narrow twins completely untwin at room temperature in approximately a day. To stimulate the untwinning process, samples with wide twin domains (Fig.4.23b) must be heated up to $T \simeq 350 \div 400$ K [4.5]. As has been shown by more thorough investigations, the untwinning is controlled by thermoactivated processes with the activation energy $U_0 = 0.59 \pm 0.05$ eV

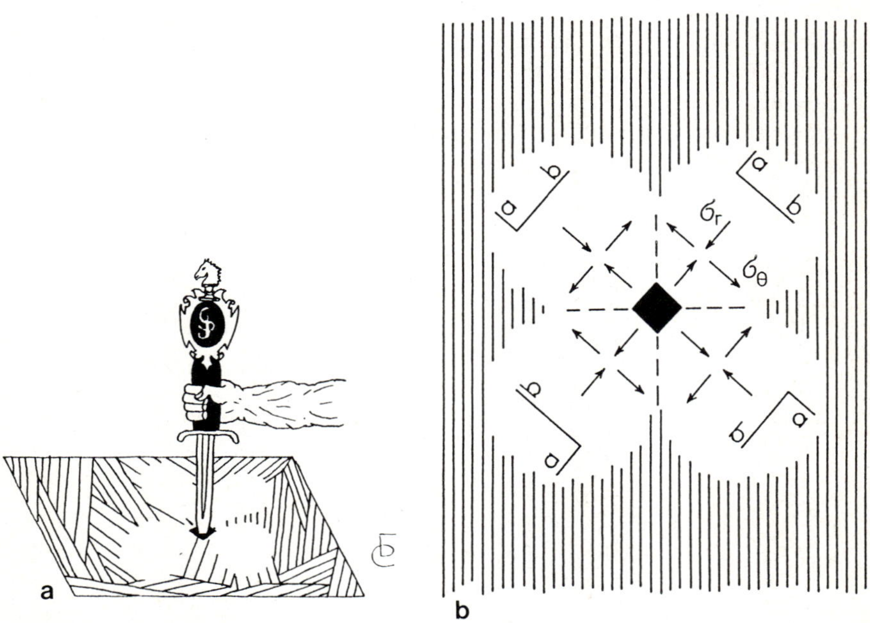


**Fig.4.24.** Illustration (**a**) and schematic (**b**) of the untwinning effect: $\sigma$ are actual stresses, **a** and **b** are crystallographic axes in the basal *ab*-plane

[4.54] (see also Chap.6). This estimate is much lower than the values of $U_0$ resulting from the microhardness temperature dependence (Tables 4.1,2). For Y-Ba-Cu-O single crystals saturated with oxygen, these estimates differ by more than a factor of 2.5. This means that deformation and alternation of the twin structure are not governed by the same processes. For instance, deformation may be associated with the generation and a motion of perfect dislocations, whereas the displacement of twin boundariues may be due to that of partial dislocations [4.27,54].

It has been noted in Sect.4.2.5 that twin restructuring cannot provide noticeable deformation of HTSCs. Upon indentation of Y-Ba-Cu-O single crystals one can see that the dimensions of the indentation do not change even slightly after monodomainization of the structure. At the same time one must take account of the relationship of the microhardness and the twin structure of the crystals [4.5]. The values of microhardness and the degree of its anisotropy were much higher for the regions with wide twin domains. The crystals with high density of narrow twins are much "softer", that is, in our case twins as carriers to mobile dislocations [4.25] are unlikely to be essential in deformation processes. The relationship between plastic properties and the twin structure can be more complicated. For instance, the twin structure may be connected with the oxygen content and state by some other structural factors, and they, in turn, define plastic and other physical properties of high-$T_c$ superconductors (Chap.6).

### 4.3.5 Generation of Dislocations and Twins

Indentation of a crystal gives rise to different defects and changes the structure in the area of the indentation [4.65,67]. This is completely true also for HTSC materials. In the previous section we discussed data related to untwinning. In some cases, upon indenting Y-Ba-Cu-O single crystals, the opposite was observed - generation of twins in monodomain regions of crystals (Fig.4.25). In the twin region inversion of the *a*- and *b*-axes relative to the crystal matrix is seen. The coloring of twins and of the sur-

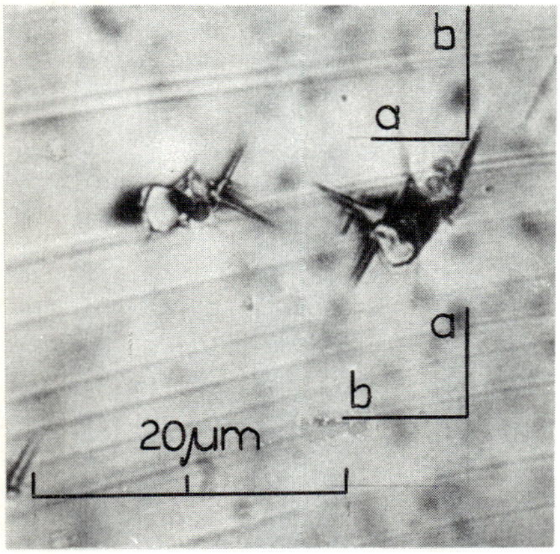


**Fig.4.25.** Twin nucleation around two indentations in the twin-free region of a $YBa_2Cu_3O_{7-x}$ single crystal

rounding crystal matrix, when the samples are rotated in crossed polaroids, varies in antiphase (bright matrix, dark twins, and vice versa). The generation of twins is related to the stress relaxation in indented crystals and is another proof of mechanical effects on the HTSC structure.

Generation of twins during indentation is a rather rare event. Generation of dislocations and their clusters is more typical in indentation and other ways of deformation [4.9, 12, 13, 24-27, 53] (Sect.4.2.5). The most detailed TEM studies of dislocations in indented Y-Ba-Cu-O single crystals have been performed by *Yoshida* et al. [4.27] who discovered that indentation at elevated temperatures leads to the generation of dislocations with the ⟨100⟩ Burgers vector. Along with these, ⟨110⟩ dislocations are generated too, which, as *Yoshida* et al. believed, arise as a result of dislocation reactions. They also have shown that ⟨100⟩ dislocations dissociate into partial dislocations with $\frac{1}{2}\langle 100\rangle$ Burgers vectors. A more detailed analysis of the dislocation structures arising under different indentation conditions is still to be carried out.

### 4.3.6 Crack Formation and Parameters of Brittle Fracture

Besides plastic deformation, cracks are also formed during indentation of ceramics and HTSC single crystals [4.4-7, 53, 61-64]. Under small loadings cracks are formed only in the corners of the indentation and do not produce any noticeable effect on $H_v$ (Sect.4.2). This makes it possible to measure simultaneously the microhardness and the critical coefficient $K_{1c}$ of stress intensities at crack tips. This coefficient characterizes the resistance to cracks in the process of brittle-viscous fracture [4.75]. As applied to HTSCs, the problem has been considered in [4.4, 6, 64]. We shall regard here the data obtained by *Demirskii* et al. [4.6a] who studied indentation of Y-Ba-Cu-O single crystals. They estimated the coefficient $K_{1c}$ from the semi-empirical relation [4.75]:

$$K_{1c} = 0.48(2d)^2 H_v/\Phi(c)^{3/2} \, , \tag{4.5}$$

where c and 2d are the dimensions of cracks and indenter patterns (Fig.4.13), $\Phi = H_v/\sigma_c \simeq 3$. Using (4.1) and the threshold loadings $P_c$ involved in crack formation, this relation can be written as

$$K_{1c} c^{3/2} \simeq 0.0752(P - P_c) \, . \tag{4.6}$$

The data presented in Fig.4.26 agree well with this dependence and yield $K_{1c} \simeq 0.4\,\mathrm{MPa\,m}^{1/2}$ [4.6a]. This estimate, as well as the values of $H_v$ given in Table 4.1 show that Y-Ba-Cu-O single crystals belong to the class of hard and brittle materials [4.65-67].

Formation of cracks in ceramic samples proceeds at loadings on the indenter lower than in single crystals. Thus, along with cracks, inside crystallites grain-boundary cracks arise, too. Such fracture is typical of ceramics materials [4.18] and we have already pointed out that grain boundaries affect the processes of HTSC ceramics fracture (Sect.4.2.1, 5).

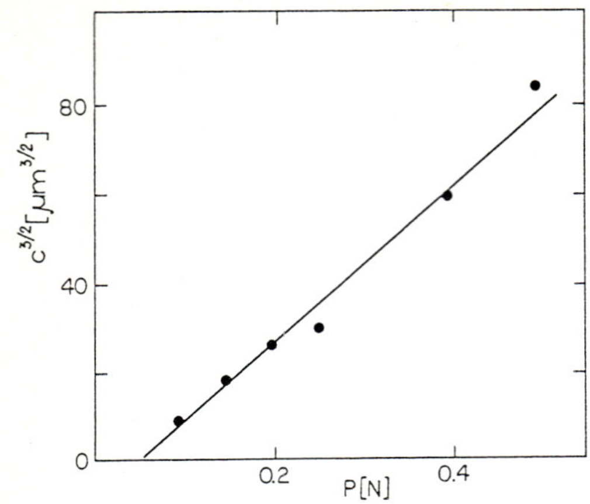


**Fig.4.26.** Length of radial crack induced by indenting a $YBa_2Cu_3O_{7-x}$ single crystal as a function of the indentor loading P

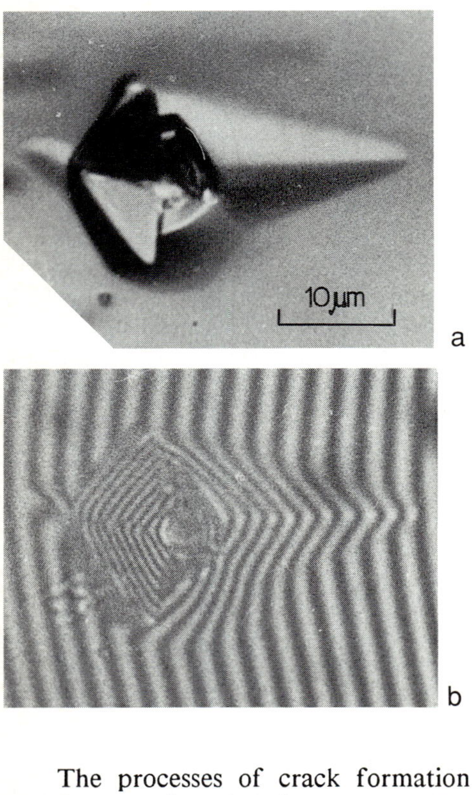


**Fig.4.27.** Lamination of the surface during indentation of a Bi-Sr-Ca-Cu-O crystal: (**a**) Photograph under oblique illumination; (**b**) interference picture

The processes of crack formation depend on the conditions of the HTSC synthesis and their structure. Here, we will only emphasize the effect of the oxygen factor on these processes. For instance, the annealing of as-grown Y-Ba-Cu-O single crystals in oxygen results in both a change in microhardness (Sect.4.3.3) and an effect on the processes of brittle fracture [4.5]. Upon indentation the initial samples exhibit the formation of radial and circular cracks. After annealing at $T_a \simeq 540$ K, when the minimum

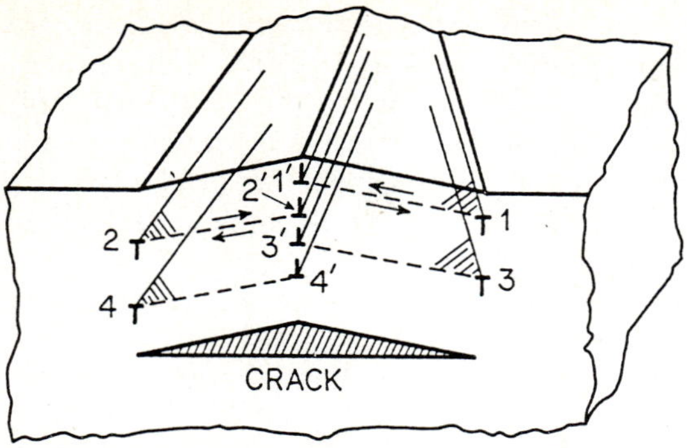


**Fig.4.28.** Scheme of deformation in crystal lamination with formation of a "roof" (Fig.4.27); the formation of three tilt boundaries by gliding of basal dislocation loops in opposite directions

was achieved for $H_V$ (Fig.4.19) no radial cracks appeared. Crack formation developed again after annealings at elevated temperatures, resulting in an increase in $H_V$. In Sect.4.3.3 we also considered the effect of oxygen content on the processes of crack formation and noted the dependence of this effect on the testing temperature [4.53a].

Finally, it should be emphasized that indentation of ceramics and HTSC single crystals may be accompanied by different processes of crack formation. For instance, in the formation of circular and radial cracks, lamination of material was observed too, as was the case in indentation of Bi-Sr-Ca-Cu-O single crystals [4.63]. Lamination elevates the material surface, which in some cases extended to a rather long distance in the form of a "roof" (Fig.4.27a). According to [4.63] this may demonstrate the essential plasticity of Bi-Cr-Ca-Cu-O single crystals. For this particular case, a calculation made on the basis of the interference pattern shown in Fig.4.27b, estimates deformation by bending as $\epsilon \geq 1\%$. A scheme of the deformation was proposed for the formation of such a relief, which is based on the building up of tilt boundaries upon motion of counter dislocation loops in the basal plane. This assumption is certainly hypothetical, but we depict this in a diagram in Fig.4.28, as requested by *Lubenets* et al. [4.63]. With this we conclude the discussion of studies of the mechanical properties of HTSC ceramics and single crystals by the method of indentation.

## 4.4 The Structure and Properties of High-$T_c$ Superconductors

Defects of the crystalline structure arising upon deformation may affect various physical properties of materials. In classic superconductors the influence of defects is principally related to a change in the Ginzburg-Landau parameter and in pinning conditions of magnetic flux vortices [4.76-79]. Deformation is known, e.g., as an element of the thermomechanical treatment to increase the critical current density $J_c$ [4.76].

Critical values of $T_c$ and of the thermodynamic magnetic field $H_c$ are fundamental parameters for superconductors [4.77-79]. The influence of defects on these parameters in the classical sense is insignificant, except for low-temperature deformation twinning which increases $T_c$ and $H_c$ [4.80-84]. Twins are a typical defect of high-$T_c$ superconductors [4.71]. For this reason twins, in particular, attracted the attention of researchers concerned with the influence of defects on the properties of these materials [4.84-101]. Investigations dealing with the effect of dislocations and grain boundaries on the properties of HTSCs are known [4.102-107].

One of the most pressing problems, whose solution is required for technical application of novel superconductors, is the production of HTSC materials with high critical current density. We have already noted the role of structure defects acting as magnetic flux pinning centers (see also Chaps.5, 6). But defects may well restrict the current-carrying ability of high-$T_c$ superconductors. Interfaces (e.g., grain boundaries in ceramics and films) play a specific role here. Note, e.g., the results of *Dimos* et al. [4.107] who observed a drastic drop of the critical current density in a bicrystalline Y-Ba-Cu-O film. An interface acts as a Josephson junction and limits $J_c$ here [4.108].

In contrast to traditional superconductors, the properties of HTSCs may be more sensitive to their structural state. Note, e.g., the dependence of Y-Ba-Cu-O properties on the concentration and position of oxygen atoms in the crystal lattice [4.51, 70, 71, 109-112]. See also the data for Bi(Pb)-Sr-Ca-Cu-O and Tl-Ba-Ca-Cu-O [4.113, 114]. Oxygen reconstruction in the region of dislocations, twins and grain boundaries all enhance the influence of these defects on the properties of HTSCs.

The problems connected with the influence of crystalline-structure defects on the properties of high-$T_c$ superconductors deserve a separate discussion. The scope of this chapter does not allow us to carry out such a discussion at sufficient depth. We shall, therefore, briefly treat only two examples illustrating the evolution of research in this field.

### 4.4.1 Deformation and Properties of Y-Ba-Cu-O Ceramics

Deformation of HTSC ceramics and single crystals leads to a change in their structure and, therefore, their properties. Inasmuch as HTSC materials are very brittle, the problems concerned with the formation of cracks and disturbances in the grain-boundary region are important for the technology of HTSCs. We shall briefly discuss the change of the current-carrying ability of Y-Ba-Cu-O ceramics as the intergrain contacts are broken by mechanical treatment under high pressure [4.57]. Investigations have shown that in this case samples may be noticeably deformed without being visibly damaged even at room temperature [4.25].

Compacting of the pressurized samples is accompanied by a shift of crystallites and destruction of intercrystallite bonds. This, in turn, leads to a drastic increase of the electric resistivity and decrease of $J_c$ [4.57]. Figure 4.29 shows the temperature dependence of $J_c$ of $YBa_2Cu_3O_{6.95}$ ceramics in

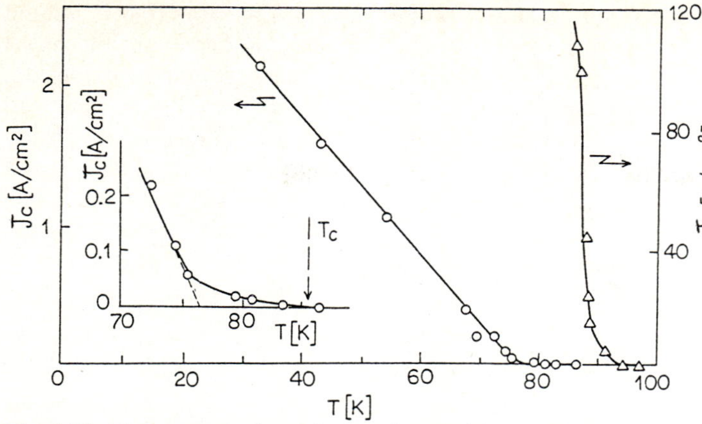


**Fig.4.29.** Temperature dependence of the critical current density $J_c$ of $YBa_2Cu_3O_{6.95}$ ceramics: Prior to the mechanical treatment (*triangles*); after all-round pressurization at $T \simeq 300$ K and $P \simeq 4.5$ GPa (*circles*); (*insert*, sample after treatment, plotted on an expanded scale)

the initial state and after a treatment at 300 K and a pressure of 4.5 GPa. The initial and the treated samples exhibited a dependence near $T_c$ given by

$$J_c = J_{co}(1 - T/T_c)^m , \tag{4.7}$$

where $m \simeq 1.5 \div 2$, and at lower temperatures this dependence becomes linear ($m \simeq 1$). It is known that (4.7) with $m \simeq 2$ indicates that the current passes through Josephson junctions [4.108]. The data suggest that the destruction of the intergrain contacts leads to worsening of the current-carrying ability of these junctions; $J_c$ drops by more than two orders of magnitude, and the temperature range in which $J_c(T)$ has the form of (4.7) broadens noticeably. These results were discussed in more detail by *Antonov* et al. [4.57]. Other reports are concerned with the influence of stress on the critical current of HTSCs. Note, for example, the results by *Dolgin* et al. [4.115] who conducted these studies at low temperatures.

Recall again that a change of the phase composition under a high temperature deformation of Y-Ba-Cu-O (Sect.4.2.2), namely a decrease of the superconducting phase concentrations along with an oxygen loss due to deformation, also leads to degradation of HTSCs [4.9, 57].

The values of $J_c$ and $T_c$ of deformed samples are completely or partly restored after annealing in an oxygen medium. Moreover, the deformation (or treatment under pressure) and thermal treatment in oxygen combined lead to a noticeable increase of $J_c$. *Wenk* et al. [4.116] observed texturing (preferential orientation of crystallites) of Y-Ba-Cu-O ceramics after a uniaxial compression at elevated temperature. After prolonged annealing of these samples in oxygen their current-carrying ability increased noticeably, i.e., annealing in oxygen is a necessary element for restoration and improvement of the properties of deformed Y-Ba-Cu-O and other HTSCs.

The principal change of HTSC properties upon deformation is related to the destruction of grain boundaries, crack formation, change in the phase composition of materials, oxygen concentration and oxygen distribution. The influence of other structure defects such as twins, dislocations and point defects is masked in ceramic samples by these stronger factors. To investigate the properties of high-$T_c$ superconductors related to twins and dislocations it is more expedient to use single srystals. Unfortunately, because of the absence of sufficiently large and high-quality single crystals, we know of only a few attempts to study the influence of deformation on the properties of such samples [4.117].

### 4.4.2 Twins and Superconductivity

Traditional superconductors containing twins of different genesis have increased critical values of $T_c$ and $H_c$ [4.80-84, 118-120] and the review [4.121]. To complete this chapter we shall briefly discuss some aspects associated with twins in HTSCs and compare the data on the temperature dependence of the critical current for Y-Ba-Cu-O and Nb single crystals.

While studying the temperature dependence of the critical current of Y-Ba-Cu-O ceramics, *Bobrov* and *Lebyodkin* [4.84] found that near $T_c$ it has the form of (4.7) with $m \simeq 1.5$-$2$ and is qualitatively similar to the data for Nb deformed by twinning. *Watanabe* et al. [4.100] and *Ogale* et al. [4.101] observed analogous dependences for Y-Ba-Cu-O ceramics and films. In the case of ceramics and films besides twins there may be other factors giving rise to singularities of $J_c(T)$, e.g., the existence of Josephson junctions in the grain boundary region [4.108] or, in the case of films, dimensionality [4.77]. We, therefore, used single crystals in our further experiments [4.84b].

In most Y-Ba-Cu-O single crystals used in [4.84] twins formed a system of intersecting domains with $\langle 110 \rangle$ directions. In the as-grown single

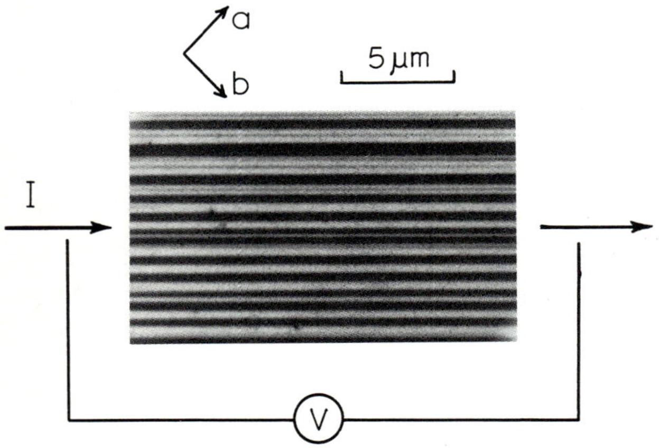


**Fig.4.30.** Optical polarizing image of a surface region of a single crystal of $YBa_2Cu_3O_{7-x}$. The direction of the electric current and schematic presentation of its measurement are shown

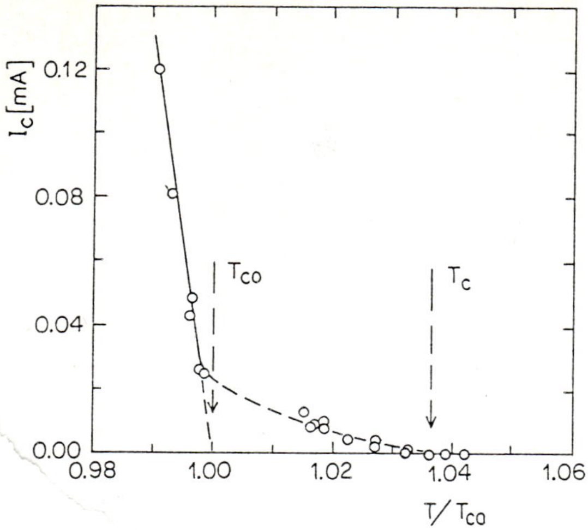


**Fig.4.31.** Normalized temperature dependence of the critical current $I_c$ of the $YBa_2Cu_3O_{7-x}$ single crystal shown in Fig.4.30

crystals the superconducting transitions occurred in the range $T \simeq 70 \div 90$ K (after annealing in oxygen, $T_c \simeq 90 \div 93$ K). Near $T_c$ the temperature dependence of the critical current $I_c$ had the form of (4.7) with $m \simeq 1.3 \div 2$ for all the investigated crystals. Now we present the data for an as-grown sample with a twinned structure in the form of parallel domains with a period of $\sim 1$ $\mu$m, exemplified in Fig.4.30. Figure 4.31 shows $I_c(T)$ at passing of the electric current domains in this sample. These data plotted on a logarithmic scale (Fig.4.33) yield $m \simeq 1 \div 4 \pm 0.1$. For comparison we present the data for a Nb single crystal in the initial and deformed states (Fig.4.32) [4.84a]. In twinned Nb $T_c$ is increased, and the dependence $J_c \sim H_c(T) \sim$

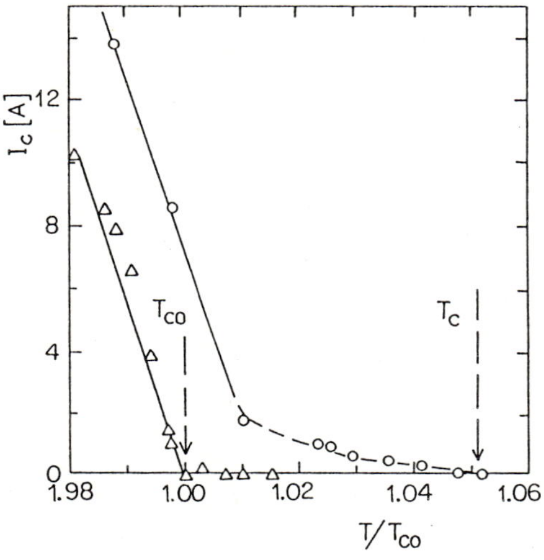


**Fig.4.32.** Normalized temperature dependence of the critical current $I_c$ of a Nb single crystal: initial sample (*triangles*); after low temperature twinning (*circles*) ($T_{co}$ is the critical temperature of the starting sample)

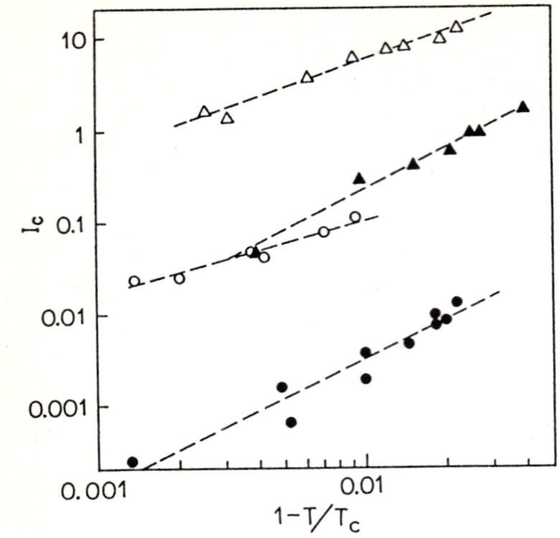


**Fig.4.33.** Data of Figs 4.31 and 32 on a logarithmic scale: *1* and *2* represent single-crystal $YBa_2Cu_3O_{7-x}$; *3* and *4* give data of Fig.4.32 on starting and deformed Nb single crystal; *light* data in the range $T \le T_{co}$, T normalized to $T_{co}$, *dark* $T_{co} \le T \le T_c$, T normalized to $T_c$

$(1-T/T_c)$, corresponding to the Silsbee rule [4.76] for bulk superconductors, is replaced near $T_c$ by a power dependence with $m \simeq 3/2$ (Fig.4.33).

A comparison of the data for Y-Ba-Cu-O and Nb single crystals shows that near $T_c$ the $J_c(T)$ are qualitatively similar, which is true in approximately identical ranges of relative temperature variation ($\Delta T/T_c \simeq 2\div5\%$). In traditional superconductors the behavior of $J_c(T)$ near $T_c$ may be related to the existence in the twin region of localized superconducting states [4.84, 121]. Possibly the same situation is observable in Y-Ba-Cu-O single crystals. This analogy between the data for classic and high-$T_c$ superconductors is quite interesting. Other factors that may affect $\mathbf{J}_c(T)$ in HTSC single crystals are the anisotropy of HTSC properties, fluctuation effects, presence of weak-link elements and an inhomogeneity in the composition. The latter must be taken into account in discussing the data for as-grown single crystals, as in [4.122] and Chap.6.

In considering the influence of twins on the HTSC properties, different, often contradictory, assumptions are made. For example, *Deutscher* [4.93] suggests that in the twin region, superconductivity is suppressed. On the other hand, we have the results of *Chaudhari* et al. [4.105] who reported the observation of high critical current values in HTSC films containing a large number of twins but free from grain boundaries. A brief discussion of the influence of twins on HTSC properties may be found in Chaps.5,6. It may also be pointed out that even in the case of superconductivity enhancement along twins (like in traditional superconductors) we may have the opposite effect of superconductivity weakening and decrease of $\mathbf{J}_c$ as the current flows across twin boundaries, i.e., the existence of energy barriers in the twin region [4.91, 123]. In the general case the role of twins in the determination of the HTSC properties may be ambiguous, and investigations in this field are far from finished. We hope that in the near future new data will be obtained for HTSC single crystals and films of different compositions, with different twin structures, in particular, for homogeneous single crystals with an oriented twin structure (Fig.4.30). Of particular interest will be investigations of the properties of HTSCs under mechanical action on the twin structure [4.5, 54-56, 117].

## 4.5 Conclusion

We have presented the data which, we hope, adequately illustrate the principal directions of research on mechanical properties of HTSCs. We were not able to present all available results, moreover the situation in this field is rapidly changing and investigations are far from being completed.

Thus, all the known HTSCs are not very plastic. In the process of their deformation, dislocations are generated and the twin structure changes. This demonstrates the role of these defects in deformation. At elevated temperatures diffusional processes may contribute to deformation of HTSCs. We point out here the particular role of oxygen and its influence on various properties of HTSCs, including mechanical ones. Note that in ceramics, the main factors affecting their properties are grain boundaries, while in single crystals and films, dislocations and twins become important.

Finally, I thank all the colleagues and coauthors for the pleasant and useful cooperation and discussions. I thank Profs. B.I. Smirnov, V.V. Shpeizman, S.V. Lubenets, V.D. Natsik and Dr. L.S. Fomenko for the fruitful discussion and permission to include some of their results in this chapter. I want to express my particular gratitude to my colleagues A.N. Izotov, M.A. Lebyodkin, E.I. Korsakova, Yu.D. Midelashvily, L.I. Velyuts and E.A. Antonova for their assistance in preparing the manuscript, and my son S.V. Bobrov for some illustrations.

## References

4.1 J.G. Bednorz, K.A. Müller: Z. Physik B **64**, 189-193 (1986)

4.2 Proc. Int'l Conf. on High-Temperature Superconductors, Materials and Mechanisms of Superconductivity (Interlaken, Switzerland), ed. by J. Müller, J.L. Olsen. Physica C **153-155** (1988)

4.3 B. Raveau, C. Michel, M. Hervieu, D. Groult: *Crystal Chemistry of High-$T_c$ Superconducting Copper Oxides*, Springer Ser. Mater. Sci., Vol.15 (Springer, Berlin, Heidelberg 1991)

4.4 R.F. Cook, T.R. Dinger, D.R. Clarke: Appl. Phys. Lett. **51**, 454-459 (1987)

4.5 V.S. Bobrov, V.K. Vlasko-Vlasov, G.A. Emel'chenko, M.V. Indenbom, M.A. Lebyodkin, Yu.A. Ossipyan, V.A. Tatarchenko, B.Ya. Farber: Fiz. Tverd. Tela **31**, 93-99 (1989) [Engl. transl.: Sov. Phys.-Solid State **31**, 597-598 (1989)]

4.6 V.V. Demirskii, H.J. Kaufmann, S.V. Lubenets, V.D. Natsik, L.S. Fomenko: Fiz. Tverd. Tela **31**, 263-266 (1989) [Engl. transl.: Sov. Phys.-Solid State **31**, 1065-1067 (1989)]
V.V. Demirskii, S.V. Lubenets, V.D. Natsik, M.N. Sorin, L.S. Fomenko: Superconductivity: Phys. Chem. Tech. **3**, 84-88 (1990)

4.7 H.C. Ling, M.F. Yan: J. Appl. Phys. **64**, 1307-1311 (1988)

4.8 N.N. Peschanskaya, B.I. Smirnov, V.V. Shpeizman, P.N. Yakushev: Fiz. Tverd. Tela **30**, 3503-3505 (1988); ibid. **31**, 271-273 (1989) (in Russian)
4.9 V.V. Bobrov, I.I. Zver'kova, A.P. Ivanov, A.N. Izotov, L.A. Novomlinskii, R.K. Nikolaev, Yu.A. Ossipyan, N.S. Sidorov, V.Sh. Shekhtman: Fiz. Tverd. Tela 32, 826-828 (1990) (in Russian)
V.S. Bobrov, V.A. Goncharov, G.A. Emel'chenko, A.P. Ivanov, R.K. Nikolaev, A.N. Izotov, L.A. Novomlinskii, Yu.A. Ossipyan, N.S. Sidorov, E.V. Suvorov, V.Sh. Shekhtman, L.N. Zave'lskaya, I.I. Zver'kova: Proc. 4th Symp. on HTSC, Trieste, Italy (1990) B-p.49
4.10 Y. Nishi, N. Ninomiya, S. Moriya, F. Kanai, K. Tachikava: J. Mater. Sci. Lett. **8**, 507-508 (1989)
4.11 O.A. Kayibyshev, R.M. Imaev, M.F. Imaev: Dokl. Akad. Nauk SSSR **305**, 1120-1123 (1989) (in Russian)
4.12 A.W. von Sumberg, Nan Chen, K.C. Goretta, J.L. Routbord: J. Appl. Phys. **66**, 2079-2082 (1989)
4.13 G. Bussod, A. Pechenik, Chung-Tse Chu, B. Dunn: J. Am. Ceram. Soc. 72, 137-139 (1989)
4.14 R.W.K. Honeycombe: *The Plastic Deformation of Metals* (Edward Arnold, Ltd, London 1968)
4.15 K.N. Tu, C.C. Tsuei, S.I. Park, A. Levi: Phys. Rev. B **38**, 772-775 (1988)
K.N. Tu, N.C. Yeh, S.I. Park, C.C. Tsuei: Phys. Rev. B **38**, 5118-5121 (1988)
H. Strauven, J.P. Locquet, O.V. Verbeke, Y. Bruynseraede: Solid State Commun. **65**, 293-296 (1988)
4.16 R. Berner, H. Kronmüller: *Plastische Verformung von Einkristallen* (Springer, Berlin, Heidelberg 1965)
H. Kronmüller: *Theorie der plastischen Verformung* (Springer, Berlin, Heidelberg 1965)
4.17 J. Fridel: *Dislocation* (Pergamon, Paris 1964)
4.18 R.L. Coble, N.M. Parikh: *Frakture*, ed. by H. Liebowitz, Vol.7 (Fractures on Nonmetals and Composites) (Academic, New York 1972)
4.19 V.V. Shpeizman, N.N. Peschanskaya, B.I. Smirnov, Yu.P. Stepanov: Fiz. Tverd. Tela **31**, 105-108 (1988) (in Russian)
4.20 S.A. Gridnev, O.N. Ivanov, A.G. Luchaninov: Izv. Akad. Nauk USSR **53**, 1349-1352 (1989) (in Russian)
4.21 V.P. Soldatov, V.D. Natsik, N.M. Chaikovskaya: Fiz. Tverd. Tela **33**, 1777-1783 (1991) (in Russian)
4.22 B.I. Smirnov: private communication (1991)
N.N. Peschanskaya, B.I. Smirnov, V.V. Shpeizman: In High-$T_c$ Superconductivity, Proc. 3rd Sov. Conf., Kahr'kov, USSR (1991) Vol.3, pp.78-79
V.V. Shpeizman, B.I. Smirnov, N.N. Peschanskaya, L.K. Markov: Fiz. Tverd. Tela **33**, 2198-2204 (1991)
4.23 P.E. Reyes-Morel, X. Wu, Chen I-Wei: *Ceramic Superconductors II*, ed. by M.F. Yan (Americ. Ceram. Soc. Westerville, OH 1989) pp.590-597
4.24 T. Muraset, K. Kuroda, K. Suzuki, H. Saka: Phil. Mag. A **62**, 583-593 (1990)
4.25 J. Rabier, M.F. Denanot: J. Less-Common Metals **164** & **165**, 223-230 (1990); Revue Physique Appl. **25**, 55-59 (1990)
4.26 X.D. Shi, R.S. Yu, Z.Z. Wang, N.P. Ong, P.M. Chaikin: Phys. Rev. B 39, 827-830 (1989)
4.27 T. Yoshida, K. Kuroda, H. Saka: Phil. Mag. A **62**, 573-572 (1990)
4.28 V.N. Belamestnikh, O.L. Khasanov, Yu Con-Syu: Superconductivity. Phys. Chem. Tech. 2, 119-127 (1989) (in Russian)
4.29 G. Cannelli, R. Cantelli, F. Cordero, G.A. Costa, M. Ferretti, G.L. Olcese: Phys. Rev. B **36**, 8907-8909 (1987)

4.30 V.D. Natsik, P.P. Pal-Val: Fiz. Nizk. Temper. **16**, 806-808 (1990) (in Russian)
P.P. Pal-Val, V.D. Natsik, L.N. Pal-Val, V.I. Dotsenko, H.-J. Kaufman: Fiz. Nizk. Temper. **14**, 1296-1301 (1988) (in Russian)
4.31 N.V. Kobelev, R.K. Nikolaev, N.S. Sidorov, Ja.M. Soifer: Phys. Status Solidi (a) **127**, 355-362 (1991)
4.32 Cheng Xiaohua, Sun Linhai, Wang Yening, Shen Huimin, Yu Zheng: Modern Phys. Lett. B **2**, 737-742 (1988)
Yening Wang, Huimin Shen, Ming Zhu, Jin Wu: Solid State Commun. **76**, 1273-1275 (1990)
4.33 M.P. Staines, J.L. Tallon, W.H. Robinson, N.E. Flower: Appl. Phys. Lett. **53**, 1560-1562 (1988)
4.34 V.I. Makarov, N.V. Zavaritskii, V.S. Klochko, A.P. Voronov, V.F. Tkachenko: J. Expt. Theor. Phys. Lett. **48**, 326-329 (1988) (in Russian)
4.35 L.T. Tsimbal, A.G. Ivanov, L.P. Volkova, A.N. Cherkasov, Yu.F. Revenko: Superconductivity. Phys. Chem. Tech. **2**, 9-12 (1989) (in Russian)
4.36 P. Lemmens, Ch. Hünnkers, Ph. Fröning, S. Ewert, H. Passing, G. Marbach, A. Comberg: J. Less-Common Metals **164** & **165**, 1129-1135 (1990)
4.37 L.G. Mansurova, K.S. Pigal'skii, V.P. Sakun, A.I. Shushin, L.G. Tsherbakova: J. Expt. Theor. Phys. **98**, 978-988 (1990) (in Russian)
4.38 G. Cannelli, R. Cantelli, F. Cordero, M. Ferretti, L. Verdini: Phys. Rev. B **42**, 7925-7930 (1990)
4.39 Y. Mi, S. Sathish, R. Schaller, W. Benoit, H. Berger: J. Less-Common Metals **164** & **165**, 1420-1427 (1990); ibid: Physica C **172**, 407-412 (1991)
4.40 H.-J. Kaufman, J. Fickert, V.D. Natsik, P.P. Pal-Val, W, Matz, K. Rudolph, K. Teske: Phys. Status Solidi (a) **123**, K11-K115 (1991)
V.D. Natsik, P.P. Pal'Val, J. Entert, H.-J. Kaufman, K. Rudolph, W. Matz: Fiz. Nizk. Temper. **16**, 293-299 (1990) (in Russian)
4.41 V.I. Startsev, V.Ya.Il'ichev, V.V. Pustovalov: *Plastichnost' i Prochnost' Metallov i Splavov pri nizkikh Temperaturakh* (Sov. Metallurgiya, Moscow 1975) pp. 163-222 (in Russian)
4.42 Yu.A. Ossipyan, V.S. Bobrov: Cryst. Res. Techol. **19**, 827-838 (1984)
4.43 V.I. Startsev: Dislocations and strength of metals at very low temperature, in *Dislocation in Solids*, ed. by F.R.N. Nabarro (North-Holland, New York 1983) Vol.6, Chap.28 pp.143-233
4.44 M.O. Eatough, D.S. Ginley, B. Morosin, E.L. Benturini: Appl. Phys. Lett. **51**, 367-370 (1987)
4.45 Yu.A. Ossipyan, R.K. Nikolaev, N.S. Sidorov, V.S. Bobrov, V.S. Tsoi: J. Expt. Theor. Phys. **47**, 310-314 (1988) (in Russian)
V.S. Bobrov, E.A. Brener, S.E. Esipov, A.N. Isotov, Yu.A. Koval', V.V. Korshunov, R.K. Nikolaev, L.A. Novomlinskii, Yu.A. Ossipyan, N.S. Sidorov, V.Sh. Shekhtman: Advanced Materials (Symp.A2), Proc. Int'l Conf. ICAM 91, Strassburg, France (1991)
4.46 G.Yu. Logvenov: private communication (1990)
4.47 V.A. Tatarchenko, G.A. Emel'chenko, N.V. Abrosimov, V.A. Borodin, L.Ya. Vinnikov, O.V. Zharikov, A.A. Zhokhov, P.A. Kononovich, V.M. Masalov, I.M. Romanenko, V.V. Ryazanov, L.I. Chernyshova: Int. J. Modern Phys. **3**, 239-302 (1989)
4.48 G.A. Emel'chenko, N.V. Abrosimov, A.V. Bazhenov, V.M. Masalov, A.A. Zhokhov, P.A. Kononovich, G.Yu. Logvenov, S.S. Khasanov: Mater. Lett. **3**, 96-100 (1990)
4.49 G.A. Emel'chenko: private communication (1991)
4.50 J.D. Jorgensen, M.A. Beno, D.G. Hinks, L. Soderholm, K.J. Volin, R.L. Hitterman, J.D. Grace, I.K. Schuller, C.U. Serge, K. Zhang, M.S. Kleefich: Phys. Rev. B **36**, 3608-3616 (1987)

4.51 R.J. Cava, B. Batlogg, C.H. Chen, E.A. Rietman, S.M. Zahurak, D. Werder: Phys. Rev. B **36**, 5719-5722 (1987)
4.52 Xi Chen, Chao Ding, Yu Wang, Xian-Ting Zen, Bao-Sheng Luo: Phys. Stat. Solidi (a) **117**, 335-340 (1990)
4.53 L.S. Fomenko, B.Ya. Farber, N.S. Sidorov, V.D. Natsik, S.V. Lubenets, V.I. Kulakov, Yu.L. Iunin, A.N. Izotov, G.A. Emel'chenko, V.S. Bobrov: Superconductivity. Phys. Chem. Techn **4**, Nr.12, 2394-2403 (1991) (in Russian)
V.S. Bobrov, G.A. Emel'chenko, L.S. Fomenko, A.N. Izotov, Yu.L. Iunin, M.P. Kulakov, S.V. Lubenets, V.D. Natsik, V.I. Nikitenko, Yu.A. Ossipyan: Priv. commun.
4.54 L.A. Dorosinskii, M.V. Indenbom, V.I. Nikitenko, B.Ya. Farber: J. Expt. Theor. Phys. Lett. **49**, 183-187 (1989) (in Russian)
4.55 D.L. Kaizer, F.W. Gayle, R.S. Roth, L.J. Swatzendruber: J. Mater. Res. **4**, 745-747 (1989)
4.56 V.K. Vlasko-Vlasov, L.A. Dorosinskii, M.V. Indenbom, V.I. Nikitenko, A.A. Polyansii, A.B. Antonov, Yu.M. Gusev, G.A. Emel'chenko: Superconductivity. Phys. Chem. Tech. **4**, 1100-1109 (1991) (in Russian)
4.57 V.E. Antonov, T.E. Antonova, O.I. Barkalov, V.S. Bobrov, A.M. Gromov, E.V. Kisterev, E.V. Konopleva; V.P. Korzhov, I.N. Kremenskaya, V.I. Kulakov, N.S. Sidorov: Superconductivity. Phys. Chem. Techn. **5**, No.4 (1992) (in Russian)
4.58 M.J. Kramer, L.S. Chumbley, R.W. McCallum: J. Mater. Sci. **25**, 1978-1986 (1990)
4.59 I.S. Smirnova: Phys. Stat. Solidi (a) **116**, 237-252 (1989); private communication (1991)
4.60 M.A. Chernysheva: Dokl. Akad. Nauk SSSR **74**, 247-250 (1950) (in Russian)
4.61 K.L. Keester, R.M. Housley, D.B. Marshall: J. Cryst. Growth **91**, 295-301 (1988)
4.62 L.B. Harris, F.K. Nyang: J. Mater. Sci. Lett. **7**, 801-803 (1988)
4.63 S.V. Lubenets, V.D. Natsik, M.N. Sorin, L.S. Fomenko, N.M. Chaikovskaya, X.-J. Kaufman, K. Fisher: Superconductivity. Phys. Chem. Tech. **3**, 1857-1862 (1991) (in Russian)
4.64 D.S. Grabko, Yu.S. Boyarskaya, R.P. Zhitaru, M.S. Kats, V.V. Tsurkan: Superconductivity. Phys. Chem. Tech. **2**, 67-71 (1989) (in Russian)
4.65 B.W. Mott: *Micro-indentation Hardness Testing* (Butterworths, London 1956)
4.66 Yu.S. Boyarskaya, D.Z. Grabko, M.S. Kats: *Fizika Processov Mikroindertirovaniya* (Sov. Stiintsa, Kishinev 1986) pp.148-164
4.67 V.I. Trefilov, Yu.V. Mil'man, S.A. Firstov: *Fizicheskie osnovy prochnosti tugoplavkikh metallov* (Sov. Naukova, Dumka, Kiev 1975) pp.64-80 (in Russian)
4.68 B. Kusz, R.J. Barczynski, M. Gazda, L. Murawski, O. Grovski, S. Stizza, J. Davoli: Physica C **160**, 25-29 (1989)
4.69 M.A. Obolenskii, V.P. Popov, Kh.B. Chashka, V.A. Starodub: Fiz. Nizk. Temper. **14**, 863-867 (1990) (in Russian)
4.70 L.T. Will, A. Berera, D. de Fontaine: Phys. Rev. Lett. **60**, 1065-1068 (1988)
4.71 C. Chaillout, M.A. Alario-Franco, J.J. Capponi, J. Chenavas, P. Strobel, M. Marezio: Solid State Commun. **65**, 283-286 (1988)
4.72 T. Graft, G. Triscone, I. Muller: J. Less-Common Metals **159**, 349-361 (1990)
4.73 V.K. Vlasko-Vlasov, M.V. Indenbom, Yu.A. Ossipyan: Sov. Phys. - JETP Lett. **47**, 312-315 (1989)
4.74 M.V. Klassen-Neklyudova: *Mechanical Twinning of Crystals* (AN SSSR, Moscow 1960) pp.44-49,110-124 (in Russian)
4.75 A.G. Evans, T.R. Wilshaw: Acta. Met. **24**, 939-955 (1976)
A.G. Evans, E.A. Charles: J. Am. Ceram. Soc. **59**, 371-372 (1976)
4.76 A.M. Campbell, J.E. Evetts: *Critical Currents in Superconductors* (Taylor and Francis, London 1972)

4.77 D. Saint-James, G. Sarma: *Type II Superconductivity* (Pergamon, New York 1969)
4.78 E.A. Linton: *Superconductivity* (Methuen, London 1969)
4.79 P.G. De Gennes: *Superconductivity of Metals and Alloys* (W.A. Benjamin, New York 1966)
4.80 G.K. Baranova, V.S. Bobrov, Ch. V. Kopetskii, G.Ya. Sal'nikov: J. Expt. Theor. Phys. **50**, 132-138 (1979) [Engl. transl.: JETP **77**, 257-269 (1970]
4.81 V.S. Bobrov, S.N. Zorin: J. Expt. Theor. Phys. Lett. **40**, 1147-1150 (1984) [Engl. transl.: JETP Lett. **40**, 345-347 (1984)
V.S. Bobrov, M.A. Lebyodkin, S.N. Zorin: Physica B **165** & **166**, 1515-1516 (1990)
4.82 V.K. Aksenov, I.A. Gindin, V.I. Sokolenko, Ya.D. Starodubov: Fiz. Nizk. Temp. **11**, 93-96 (1985) (in Russian)
4.83 A.M. Dolgin: Fiz. Nizk. Temp. **13**, 1015-1025 (1987) (in Russian)
4.84 V.S. Bobrov, M.A. Lebyodkin: J. Expt. Theor. Phys. Lett. **47**, 715-719 (1988)
V.S. Bobrov, M.A. Lebyodkin: Physica C: **178**, 411-413 (1991)
4.85 A.A. Abrikosov, A.I. Buzdin, M.L. Kulik, D.A. Kuptsov: Int'l J. Phys. B **2**, 1045-1052 (1988)
A.A. Abrikosov, A.I. Buzdin: J. Expt. Theor. Phys. Lett. **46**, 463-465 (1987) (in Russian)
4.86 A.F. Andreev: J. Expt. Theor. Phys. Lett. **46**, 463-465 (1987) (in Russian)
4.87 V.L. Indenbom: J. Expt. Theor. Phys. Lett. **46**, 144-147 (1987) (in Russian)
V.L. Indenbom, E.B. Loginov: Kristallografiya **32**, 1077-1081 (1987) (in Russian)
4.88 L.N. Bulaevskii, M.V. Zyskin: Fiz. Tverd. Tela **31**, 246-250 (1989) (in Russian)
4.89 Ph. Lambin, J.-P. Vigneron, Amand A. Lucas: Physica C **153-154**, Pt.2, 1241-1242 (1988)
4.90 D. Shi, M. Xu, M.S. Boley, U. Welp: Physica C **160**, 417-423 (1989)
4.91 A.Vl. Gurevich, R.G. Mints: J. Expt. Theor. Phys. Lett. **48**, 618-621 (1988) (in Russian)
4.92 U. Welp, M. Grimsditch, H. You, W.K. Kwok, M.M. Fang, G.W. Crabtree, J.Z. Lin: Physica C **161**, 1-5 (1989)
4.93 G. Deutscher: Physica C **153-155**, Pt.2, 15-20 (1988)
4.94 N. Garcia, S. Vieira, A.M. Baro, J. Tornero, M. Pazos, L. Va'zquez, J. Gomez, A. Aguilo, S. Bourgeal, A. Buendia, M. Hortal, M.A. Lopez de la Torre, M.A. Ramos, R. Villar, K.V. Rao, D.-X. Chen, J. Nogues, N. Karpe: Z. Physik B **70**, 9-13 (1988)
4.95 K.D. Schotte, T.T. Truong: Physica C **153-155**, Pt.2, 1201-1202 (1988)
V. Sebastian, S.A. Solin, N. Garcia, M. Hortal, A. Aguilo: ibid. C **153-155**, Pt.1, 984-985 (1988)
4.96 G. Deutscher, K. Muller: Phys. Rev. Lett. **59**, 1745-1747 (1987)
4.97 A. Sobyanin, A. Stratonnikov: Physica C **153-155**, Pt.2, 1681-1682 (1988)
4.98 S.E. Inderhees, M.B. Salamon, N. Goldenfeld, J.P. Rise, B.G. Pazol, D.M. Ginsberg: Phys. Rev. **60**, 1178-1180 (1988)
4.99 M.M. Fang, V.G. Kogan, D.K. Finnemore, J.R. Clem, L.S. Chumbley, D.E. Farrel: Phys. Rev. B **37**, 2334-2337 (1988)
4.100 H. Watanabe, Y. Kasai, T. Mochuku, A. Sugishita, I. Iguchi, E. Yamaka: Jpn. J. Appl. Phys. **26**, L615-1657 (1987)
4.101 S.B. Ogale, D. Dijkkamp, T. Venkatesau, X.D. Wu, A. Inam: Phys. Rev. **36**, 7210-7213 (1987)
4.102 Ding Shiying, Yu Zheng, Chen Jun: Chin. Phys. Lett. **6**, 24-27 (1989)
4.103 S. Nakahara, S. Jin, R.C. Sherwood, T.H. Tiefel: Appl. Phys. Lett. **54**, 1926-1928 (1989)

4.104 Y. Takahashi, M. Mori, Y. Ishida: J. Jpn. Inst. Met. **53**, 477-478 (1989)
4.105 P. Chaudhari, F.K. Le Goues, A. Segmüller: Science **238**, 342-344 (1987)
4.106 D.M. Hwang, T. Venkatesan, C.C. Chang, L. Nazar, X.D. Wu, A. Inam, M.S. Hegde: Appl. Phys. Lett. **54**, 1702-1704 (1989)
4.107 D. Dimos, P. Chaudhari, J. Maunhart, F.K. Le Gones: Phys. Rev. Lett. **61**, 219-222 (1988)
P. Dimos, P. Chaudhari, J. Maunhart: Phys. Rev. B **41**, 4038-4049 (1991)
4.108 A. Barone, G. Paterno: *Physics and Applications of the Josephson Effect* (Wiley, New York 1982)
4.109 V.G. Bar'yakhtar, A.V. Zhalko-Titarenko, V.S. Mel'nikov, I.G. Mikhailov, A.V. Morozovsky, V.V. Nemoshkalenko, V.V. Pan, N.P. Pshentsova, S.K. Tolpygo: Int'l J. Mod. Phys. B 2, 1259-1266 (1988)
4.110 G. Oszlanyi, S. Pekker, Gy. Hutiray, E. Zsoldos, L. Mihaly: Int'l J. Mod. Phys. B 2, 1305-1311 (1988)
4.111 A.A. Sinchenko, N.V. Zakharchenko, G.V. Korolev, S.V. Shavkin: Superconductivity. Phys. Chem. Tech. 2, 13-16 (1989) (in Russian)
4.112 A.F. Moodie, H.J. Whitfield: Ultramicroscopy **24**, 329-338 (1988)
4.113 N. Muira, H. Suzuta, Y. Deshimaru, Y. Shimizu, H. Sakashita, N. Yamazoe: Jpn. J. Appl. Phys. **28**, 1112-1114 (1989)
4.114 A. Schilling, H.R. Ott, F. Hulliger: Physica C **157**, 144-148 (1989)
4.115 A.M. Dolgin, V.D. Natsik, S.N. Smirnov: Fiz. Nizk. Temp. **16**, 1326-1337 (1990) (in Russian)
4.116 H.R. Wenk, J. Pannetier, G. Bussod, A. Pechenik: J. Appl. Phys. **65**, 4070-4073 (1989)
4.117 V.S. Bobrov, V.K. Vlasko-Vlasov, M.V. Indenbom, M.A. Lebyodkin, B.Ya. Farber: In Physics of Low Temperatures (Superconductivity), Proc. 25th Sov. Conf., Leningrad, USSR (1988) Vol.1, pp.171-172 (in Russian)
4.118 M.S. Khaikin, I.N. Khlustikov: Pis'ma Zh. Eksp. Teor. Fiz. **33**, 167-170 (1981) [Sov. Phys. JETP Lett. **33**, 158-160 (1981)]
4.119 I.N. Khlustikov, M.S. Khaikin: J. Expt. Theor. Phys. Lett. **34**, 207-211 (1981); ibid **36**, 132-135 (1982); ibid: **38**, 191-194 (1983) (in Russian)
4.120 I.N. Khlustikov, S.I. Moskvin: J. Expt. Theor. Phys. **89**, 1846-1855 (1985) (in Russian)
4.121 I.N. Khlustikov, A.I. Busdin: Adv. Phys. **36**, 271-330 (1987)
4.122 V.K. Vlasko-Vlasov, M.V. Indenbom, V.I. Nikitenko, Yu.A. Ossipyan, A.A. Polyansii, R.L. Prozorov: Superconductivity. Phys. Chem. Tech. **3**, 1199-1213 (1990) (in Russian)
4.123 S.N. Burmistrov, L.B. Dubovski: Fiz. Tverd. Tela **26**, 1719-1724 (1984) (in Russian)

# 5. Vortex Structure in Single-Crystal High-$T_c$ Superconductors

L.Ya. Vinnikov, I.V. Grigor'eva, and L.A. Gurevich

*Seeing is believing*

The magnetic flux structure in type-II superconductors was predicted in 1957 by *Abrikosov* [5.1]. According to that theoretical work the magnetic field penetrates into a superconductor in the form of individual magnetic lines (vortices) each carrying one magnetic flux quantum $\Phi_0 = hc/2e = 2.07 \cdot 10^{-7}$ G·cm$^2$. The idea of a vortex state proved to be extremely valuable in explaining electrodynamic properties of type-II superconductors. The most convincing evidence of the validity of this theory was the direct observation of the Abrikosov vortices using the high-resolution technique of decoration with dispersed ferromagnetic particles. This technique was developed in 1966 by *Träuble* and *Essmann* in Stuttgart [5.2]. Their works marked the beginning of magnetic structure investigations in conventional low-$T_c$ type-II superconductors; the results have been summarized in reviews and monographs [5.3, 5].

It should be noted that although many different methods for magnetic-structure investigation such as neutron scattering, magneto-optical techniques and others [5.4] exist, the technique of decoration appeared to be most effective for observation of magnetic structure with individual vortex resolution. The other techniques do not provide this resolution. By using the technique of decoration it becomes possible to match up the observed local distortions in the vortex distribution to specific features of superconducting crystals, both local (for example, a pinning center) and extended (anisotropy). In addition, the obtained vortex patterns are easy to be understood, which is also important. However, this method is restricted to low magnetic fields (vortex separations $>\lambda$), gives only indirect information of the bulk vortex distribution (in contrast to the neutron-diffraction and $\mu$SR techniques), and the results are very difficult to interpret if dynamic effects are involved.

HTSCs are type-II superconductors with a large Ginzburg-Landau (GL) parameter $\kappa >> 1$. The magnetic flux distribution for such materials, as predicted by *Abrikosov*, is a regular, triangular vortex lattice provided the crystal is isotropic and perfect, free of defects and inhomogeneities. In fact, the regular Flux Line Lattice (FLL) has been observed in specially prepared, perfect low-$T_c$ crystals by different methods [5.5, 6]. However, real superconducting materials are in many aspects different from ideal ones. In particular, their magnetic flux structure appears to be much more complicated and, therefore, its direct observation is of great interest. As far as HTSCs are concerned, their magnetic properties display a strong anisotropy, especially in the magnetic field directions parallel and perpendicular

Springer Series in Materials Science, Vol. 23
**The Real Structure of High-T$_c$ Superconductors**
Editor: V.Sh. Shekhtman 

to the $c$-axis. X-ray and electron-microscopic investigations show crystal imperfections which are a source of irreversible behavior of the FLL, i.e., of magnetic flux pinning. All this, together with other unusual properties of high-$T_c$ superconductors, generated a new wave of interest in studying their vortex structure. This structure reflects many important characteristics of superconductors which may be helpful in understanding the superconductivity mechanism. They are the anisotropy of superconducting properties, the penetration of the magnetic field (and the penetration depth), the value of the magnetic-flux quantum, etc.

The other important aspect of the direct observation of a magnetic structure is the possibility to study vortex interaction with crystal defects - pinning effects - which are essential to possible applications. One of the interesting features of the HTSC is the extremely high upper critical field and, correspondingly, the very short coherence length, comparable with the crystal lattice parameters. This may cause considerable differences in the pinning mechanisms for HighTSC and LowTSC, in particular, a much higher effectiveness of the atomic scale crystal defects such as vacancies, stacking faults, interstititals, dislocations, etc.

The purpose of this chapter is to summarize both our results and the results of other researchers on direct observations of HTSC vortex structure by means of the decoration technique. In the following we shall consider the physical basis of the decoration method, basic properties of the vortex lattice, and the anisotropy and pinning effects in HTSC.

## 5.1 Sample Preparation and Experimental Technique

As mentioned above, the technique of decoration with dispersed ferromagnetic particles was first developed by *Träuble* and *Essmann* for the observation of the magnetic structure of superconductors with high resolution, allowing visualization of the vortex lattice in type-II superconductors [5.7]. Our technique [5.8] differs from theirs by the use of a decoration chamber instead of an open volume with evacuated helium gas.

The key problem of the decoration technique is how to prepare dispersed ferromagnetic particles $50 \div 100$ Å in size during the decoration process and how to transport them to a superconductor's surface without them sticking together. The latter causes a dramatic loss of resolution. The necessary size and concentration of ferromagnetic particles are obtained by evaporating Fe, Ni or another ferromagnetic material in a background of inert gas (helium) under a pressure of 0.1 torr at 4.2 K. If the temperature of the experiment differs from 4.2 K then the pressure should be varied until optimum particle sizes are obtained. Figure 5.1 illustrates the method. Evaporated iron particles are drawn into regions of strong magnetic field gradients at the superconductor surface, which appear in the vicinity of the vortex exit (entrance) or at any other inhomogeneity of the magnetic field.

We used single-crystal samples of the 1-2-3 (Y-Ba-Cu-O, Ho-Ba-Cu-O) and 2-2-1-2 (Tl-Ba-Ca-Cu-O, Bi-Sr-Ca-Cu-O) systems. Our attempts

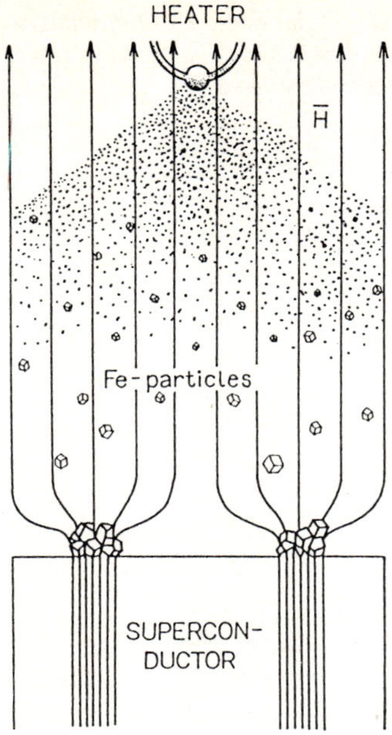


**Fig.5.1.** Sketch illustrating the main idea of the decorating method: Iron particles fly along the magnetic-field gradient and fall to the points of vortex outlets decorating them

to study the magnetic structure of ceramics were unsuccessful, as was the case for *Ourmazd* et al. [5.9]. As a rule, the investigated surface was optically smooth and was not subjected to any further processing. The crystals were grown by slow cooling of a molten mixture of the corresponding oxides and carbonates in nonstoichiometric composition [5.10]. Samples were platelets approximately $1\times1$ $mm^2$ and several tens of microns thick. The orientation of the broad surface was perpendicular to the $c$-axis. During a decoration experiment the sample was cooled in a magnetic field from room temperature to the decoration temperature ($4.2\div20$K) - the field cooling regime. The external magnetic fields were kept to <100 Oe. Higher fields worsened the resolution when the magnetic field of vortices overlap. It was possible to study the vortex structure of high-$T_c$ crystals in fields with $H_e \ll H_{c1}$ since the demagnetization factor for flat platelets with $H_e \| c$ is close to 1. The obtained decoration patterns were examined in a Scanning Electron Microscope (SEM) or an optical microscope. In SEM the vortices are observed as conglomerates of dispersed ferromagnetic particles - bright spots against a dark background. In the optical microscope the contrast is the opposite.

## 5.2 Characteristics of the Vortex Structure

Figure 5.2 shows a triangular FLL in the most perfect $Tl_2Ba_2Ca_1Cu_2O_x$ single crystal that we could obtain. For the first time the FLL (slightly disordered) has been observed in Y-Ba-Cu-O single crystals. From a comparison of the vortex number per unit area, n, and the inductance B (the

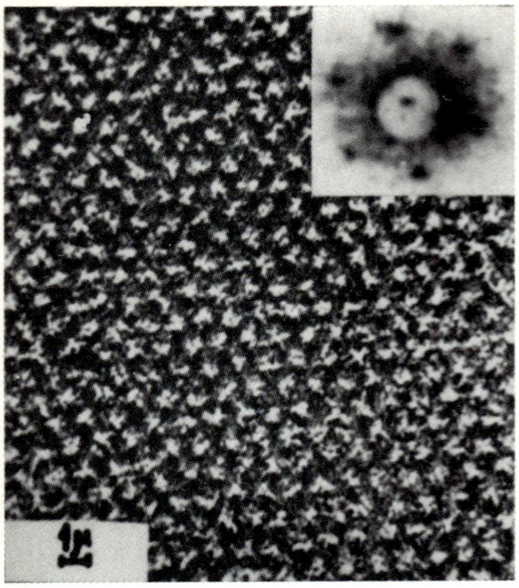

**Fig.5.2.** Regular triangular flux line lattice in the most perfect $Tl_2Ba_2Ca_1Cu_2O_x$ single crystal that we could obtain

latter was measured by a Hall probe placed directly under the crystal) the flux value per single vortex, $\Phi_1$, was measured as being practically equal to $\Phi_0 = hc/2d = 2\cdot10^{-7}\,G\cdot cm^2$. This result, found for all investigated families of superconductors, testifies a general fundamental property - the current transfer by pairs of carriers [5.11-13].

Under optimum decoration conditions the dimensions of the ferromagnetic-particle agglomerations in the decoration patterns (at least in low fields) match the magnetic-field penetration depth $\lambda$ because the particles are collected near the maximum magnetic field gradient around the vortex core. Thus, at the sample surface this region is about $2\lambda(T)$ in size. Disregarding some uncertainty in finding the "vortex diameter" we estimated $\lambda(4.2K)$ to be as less than 0.2 $\mu$m for the Y-Ba-Cu-O *ab*-plane [5.14].

The regularity of a vortex lattice of a superconductor can be disturbed not only by crystal defects but also by elastic properties of the FLL depending on external parameters, such as the magnetic-field value and temperature. Thus, as the observations in [5.15] demonstrated, the FLL defects in the same single domain of a Y-Ba-Cu-O crystal disappear upon increasing the external magnetic field. In contrast, increasing the temperature of decoration in a given magnetic field causes a disturbance of the long range order in the FLL in perfect Bi-Ca-Sr-Cu-O single crystals [5.16].

Recently, a number of theoretical works have appeared which predict the existence of topological phase transitions of the "order-disorder" type in the vortex structure of high-$T_c$ superconductors. Apart from the well-known Abrikosov lattice, new phases are expected in certain ranges of field and temperature, such as the vortex liquid (melted vortex lattice) [5.17], hexatic phase, entangled and disentangled vortex liquid [5.18].

Direct observations of the FLL by means of the decoration technique at different temperatures allow detection of a transition from an ordered structure to a disordered one. But, unfortunately, the cause of this transition is not likely to be understood from decoration patterns. In principle, with

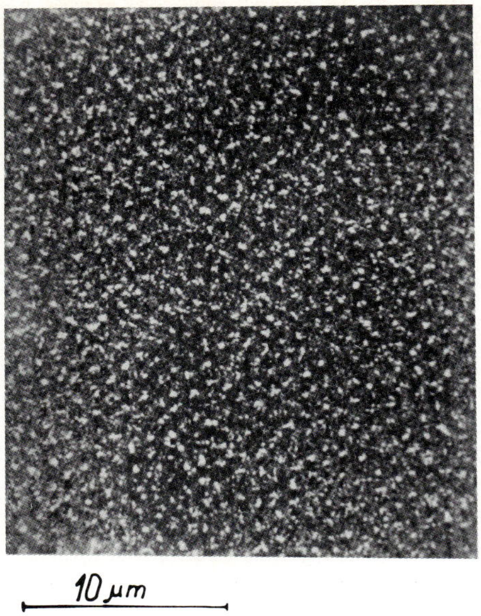


**Fig.5.3.** FLL image in $Tl_2Ba_2Ca_1Cu_2O_X$ crystal at 20 K

this technique one can see the FLL melting but not distinguish entangled and disentangled phases. Thus, *Kleinman* et al. [5.16] examined in the same decoration experiment the vortex patterns in Y-Ba-Cu-O and Bi-Sr-Ca-Cu-O single crystals at 15 K. While in Y-Ba-Cu-O the FLL did not differ considerably from that at 4.2 K, in Bi-Sr-Ca-Cu-O the regular FLL at 4.2 K was found to be destroyed and vortex images were smeared out. The latter result is interpreted as evidence of vortex motion during the decoration process. However, our experiments of the same type as in [5.16] on Y-Ba-Cu-O and perfect Tl-Ca-Ba-Cu-O single crystals did not give this result. Figure 5.3 depicts the FLL image in a Tl-Ca-Ba-Cu-O crystal at 20 K. The long-range (orientational) order is seen to be preserved at distances of several dozens of vortex spacings. The results of decoration of Y-Ba-Cu-O crystals are also analogous to those at 4.2 K. A possible reason for the latter result might be a higher (compared to Bi crystal) melting temperature of the FLL in Tl crystals. Unfortunately, we failed to observe decoration patterns at temperatures higher than 20 K.

A detailed examination of the FLL topology in perfect Bi-Sr-Ca-Cu-O crystals is given in [5.19]. Using a statistical analysis for the translational (positional) and orientational order of the vortex lattice it was found that the former disappears exponentially at a correlation length of about several vortex spacings while the latter is retained at distances of hundreds of vortex separations following a power law with the index 0.06. Through these results *Murray* et al. concluded that the vortex structure in these materials should not be treated as a regular lattice but rather as a vortex glass in a hexatic phase. They considered two possible versions of formation of this structure. One of them is freezing ("quenching") the disorder or the distortions in the vortex structure, which are characteristic of the hexatic liquid state existing at high temperatures. The other assumes that the structure observed on the decoration patterns corresponds to that which is stable at the temperature of the experiment and which appears as a result of random pinning potentials. At present, it is hardly possible to give preference to this or that model.

## 5.3 Anisotropy Effects

One of the important characteristics of high-$T_c$ superconductors, which determines a number of their unusual properties, is the strong anisotropy of superconducting parameters. It arises from the crystal structure of HTSC. As is well known, high-$T_c$ crystals consist of conducting $CuO_2$ layers which are, according to current understanding, also the superconducting planes. These layered superconducting systems can behave both as three-dimensional (3D) and two-dimensional (2D) superconductors. The dimensionality is determined by the ratio of interlayer spacing d to the correlation length $\xi$ in the direction perpendicular to the layers [5.20, 21]. If the condition $\xi(T) > d/2$ holds [5.21] then the superconducting layer coupling is relatively strong and the superconductor is three-dimensional. In the opposite case the layer coupling may occur via tunnelling of superconducting pairs, i.e., it becomes of Josephson type, and the superconductor behaves as quasi-two-dimensional. Due to the coherence length $\xi(T)$ being temperature dependent, the systems which are quasi-2D at low temperatures can undergo a 2D-3D transition with increasing the temperature to $T^*$ withe $\xi_\perp(T^*) = d/\sqrt{2}$ [5.2]. In the most anisotropic materials based on Bi and Tl such a transition is likely to occur rather close to $T_c$ so that they must behave as 2-dimensional over a wide temperature range [5.22]. This has, in fact, been confirmed by magnetization measurements [5.22] and by observations of 2D fluctuations (Kosterlitz-Thouless transition) [5.23]. For less anisotropic crystals of the 1-2-3 system 2D effects have not yet been observed but, judging by the above criteria, here a 3D-2D transition is also possible at low temperatures.

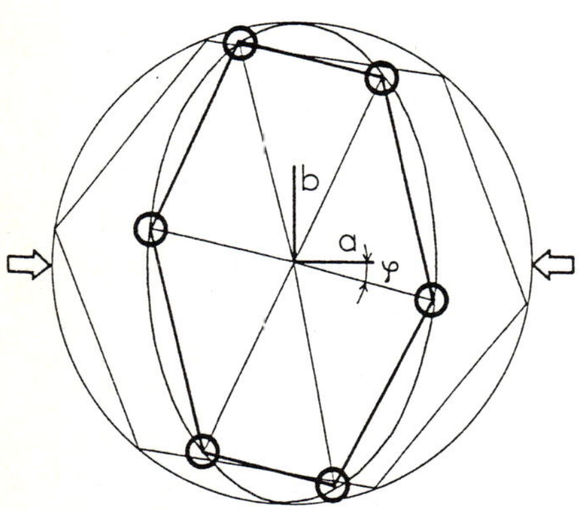


**Fig.5.4.** Compression of a regular lattice along the "heavy" axes (here, *a*-axes)

The theoretical description of quasi-2D superconducting systems is usually given within the framework of the Lawrence-Doniach model incorporating a 2D Ginzburg-Landau functional for every layer with the energy of Josephson coupling taken into account. As shown in [5.24], this description can be reduced to the 3D anisotropic GL theory if the effective mass of carriers in the direction perpendicular to the layers is defined through the Josephson coupling parameter and the interlayer spacing. Therefore, it is usually possible to describe the anisotropy effects on properties of HTSC within the anisotropic GL theory which is generalized by replacing the effective mass of the isotropic case with an effective mass tensor, i.e., different effective masses are ascribed to the carriers depending on the directions of their motion [5.25, 26]. In the system of crystallographic axes this tensor becomes diagonal:

$$M_{i,j} = \begin{pmatrix} m_a & 0 & 0 \\ 0 & m_b & 0 \\ 0 & 0 & m_c \end{pmatrix} = m_0 \hat{\mu} \qquad (5.1)$$

where $m_0 = \sqrt{m_a m_b m_c}$, $m_a$, $m_b$, and $m_c$ being the effective masses along the crystal axes. The penetration depth $\lambda$ and coherence length $\xi$ also become tensors:

$$\hat{\lambda} = \lambda_0 \hat{\mu}^{1/2} \quad \text{and} \quad \hat{\xi} = \xi_0 \hat{\mu}^{-1/2} , \qquad (5.2, 3)$$

which causes $H_{c2}$ and $H_{c1}$ to become angle dependent and, as a consequence, the vortex interaction to be anisotropic. The latter is manifested in the asymmetry of the FLL elementary cell.

The case when the field direction is along the principal axis of the anisotropy tensor (one of the crystal axes) is simplest for analysis. Abrikosov vortices lie along the magnetic field and the FLL unit cell is contracted along the axis for which the effective mass is larger (in the plane perpendicular to the vortex axis). So, if a vortex lies along the *c*-axis and $\mu_a > \mu_b$, the lattice will be contracted along the *a*-axis with a factor $(\mu_a/\mu_b)^{1/2}$ [5.26] (see the scheme in Fig.5.4). In addition, the flux tube which is usually used as an approximation for a vortex becomes oval instead of round due to the anisotropy of $\lambda$. Thus, the FLL in anisotropic superconductors is an indicator of the anisotropy and allows determination of the effective mass ratio. Experiments of this type were performed on single crystals of Y-Ba-Cu-O both on the *ab*-plane [5.27, 28] and on the plane parallel to the *c*-axis [5.28].

### 5.3.1 Vortex Lattice Anisotropy on the Basal Plane in Y-Ba-Cu-O Single Crystals

As it is well known, at low temperatures Y-Ba-Cu-O single crystals have an orthorhombic structure of the $P_{mmm}$ type [5.29] and superconducting properties should be anisotropic along all three axes. Thus, it was found that the conductivity in the normal state measured by the phase shift of reflected light [5.30], is larger in the *b*-direction than along the *a*-axis. For this reason it should be expected that anisotropy of superconducting properties exists in the *ab*-plane.

It has not been possible to determine the anistropy of $\lambda$, $H_{c1}$ and $H_{c2}$ in the *ab*-plane by means of magnetic measurements because natural single crystals of Y-Ba-Cu-O contain a number of twin domains typically << 1 $\mu$m in size so that the *a*- and *b*-directions in the crystal appear to be mixed. For this reason, decoration experiments proved to be practically the only method allowing determination of the effective mass anisotropy from the vortex patterns on the *ab*-plane.

In our study [5.27] we examined the shape of the FLL elementary cell in large monodomains which were distinguished by using a polarizing microscope. Dimensions of the domains were large ($\simeq 50 \mu$m) compared to the intervortex spacings in the magnetic fields used. The FLL images were obtained in fields of 10 and 100 Oe. In order to obtain general information about the shape of the FLL unit cell, the images of the vortex lattice were processed by a laser diffractometer. A typical example of vortex image with a corresponding diffraction pattern in shown in Fig.5.5. It is seen that the FLL is not hexagonal but slightly contracted by a factor $\eta$. This factor, along with the direction of contraction, was deduced, through a numerical calculation, from the angles $\alpha$, $\beta$ and $\gamma$ in the distorted triangular cell of the vortex lattice, measured via diffraction patterns. Investigations of the crystals in polarized light showed that for the domains with the same orientation of the *a*- and *b*-axes (such domains are identical in color when the surface of the sample is observed in polarized light) the FLL is contracted along the same direction which is perpendicular to the *b*-axis. The latter was determined with the help of a Berek phase compensator as the phase-

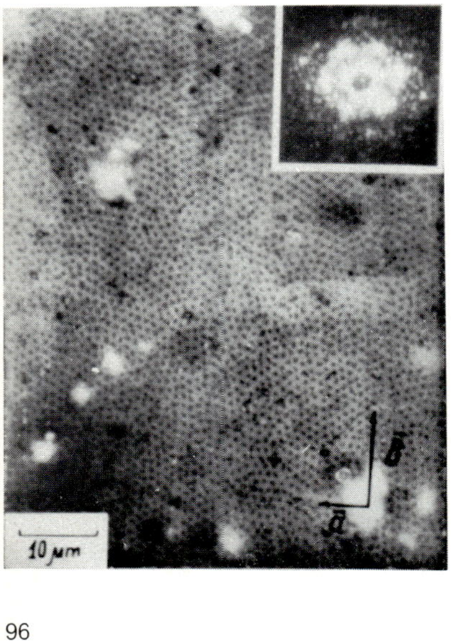


**Fig.5.5.** Vortex lattice on a basal plane in a twin-free region of a $YBa_2Cu_3O_x$ single crystal. The average magnetic induction is B $\simeq$ 10 G. The *insert* shows a diffraction pattern from this FLL image (the photograph was taken in an optical microscope)

lag axis of reflected light. Comparison of the results obtained for the external fields of 10 and 100 Oe showed that the contraction factor did not change, although the field was reduced, and was $\eta = (\mu_a/\mu_b)^{1/2} = 1.2\mp0.1$. It is interesting to note that in most cases one of the close-packed directions in the vortex lattice runs parallel to the *a*- or *b*-axis.

A similar result, also obtained with the help of the decoration technique, was obtained in [5.28]: $\eta = 1.13\mp0.02$, and later, when it became possible to prepare untwinned single-domain crystals of Y-Ba-Cu-O, the same ratio of effective masses was found from measurements of the $H_{c1}$ anisotropy in the *ab*-plane [5.31].

### 5.3.2 Flux-Line Lattice Anisotropy in the Plane Parallel to the *c*-Axis in $YBa_2Cu_3O_x$ Single Crystals

An FLL has only been observed by *Dolan* et al. in the plane parallel to the *c*-axis (due to the twinned structure of Y-Ba-Cu-O single crystals this plane is, in fact, a mixture of *ac*- and *bc*-planes). The FLL anisotropy, which could be expected in accordance with the magnetic measurements, is rather large compared to the anisotropy in the *ab*-plane. However, as noted in [5.28], the observation of individual vortices proved to be difficult because the decorating ferromagnetic particles are affected by relatively weak magnetic forces because of the very large "magnetic size" of the vortices. *Dolan* et al. stated that they could successfully observe vortices only in very low fields ($\leq$4Oe) and in a small number of experiments.

At fields comparable to the Earth's ambient field *Dolan* et al. observed isolated, oval-shaped vortices with a ratio of the half-axes (of a vortex image) of 2÷4. This oval shape indicates a penetration depth $\lambda_{a,c}$ along the *c*-direction smaller than $\lambda_{a,b}$. Upon increasing the magnetic field up to 4 Oe, vortex ordering was observed and formation of horizontal chains in the *c*-direction (Fig.5.6). It is interesting that many of the chains were aligned with twin boundaries, which came to the crystal surface, and only those chains formed straight lines. The other chains were slightly modulated in the vertical direction perpendicular to *c* with a period of about eight vortex spacings. It is likely that this easy formation of the FLL distortions is due to a very low shear modulus for this orientation.

As the field was further increased to 8 Oe chains of decorating particles were still observed but the vortices inside the chains were not resolved. The distance between the chains changed as $\sqrt{B}$ with further increase in the field. In this orientation each vortex carries one flux quantum $\Phi_0$ as in the previous experiments in the *ab*-plane. The FLL anisotropy, determined as the ratio of the distance between horizontal vortex chains and the in-chain nearest-neighbor distance, was $\lambda_{a,b}/\lambda_{a,c} \simeq \lambda_{a,b}/\lambda_{b,c} = 5.5\mp1$. This result is in good agreement with the anisotropy found from the ratio of critical fields [5.22].

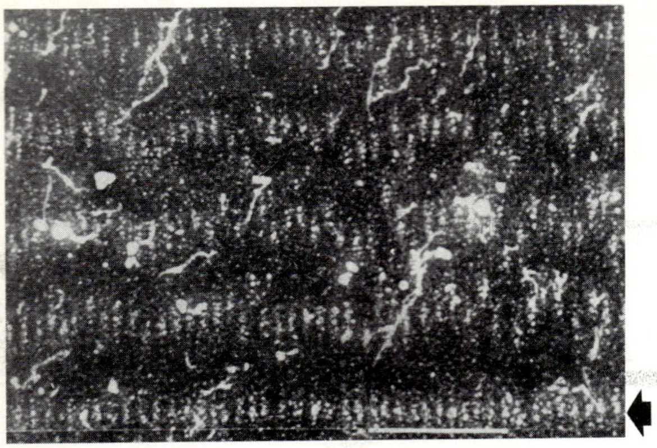

**Fig.5.6.** Horizontal chains of oval vortices at a field of 4 G perpendicular to the *c*-axis. The *arrow* indicates the position of a twin boundary which is aligned with a vortex chain. The bright, irregular objects are irrelevant debris. The *marker* is 10 $\mu$m [5.28]

### 5.3.3 Vortex Structure in Tilted Magnetic Fields

The situation when the magnetic field is tilted with respect to the *c*-axis is more complicated (the anisotropy in the basal plane is much less than the anisotropy between the *c*-axis and the *ab*-plane, so in the following we shall consider high-$T_c$ superconductors as uniaxial superconductors). The electromagnetic energy of a vortex is minimum for the field orientation parallel to $CuO_2$ layers because then $H_{c1}$ is at a minimum. For this reason it becomes advantageous for a vortex to incline away from the direction of the external magnetic field to be closer to the *ab*-plane. This deflection increases with increasing anisotropy and decreases with increasing magnetic field [because the vortex energy increases as $HB(1\text{-}\cos\theta)$]. At some combination of field and tilt angle $\theta$ the vortex should be oriented entirely in the *ab*-plane ("easy" plane $m_{ab} < m_c$ [5.32, 33]). A line of the above transition on the H-$\theta$ diagram was calculated, for example, in [5.34], where it has been referred to as a "lock-in" transition. For the quasi-2D behavior the *ab* orientation of the vortex becomes even more preferable.

As calculations have shown, in highly anisotropic superconductors the longitudinal component of the magnetic field inside a tilted vortex should change its sign in the plane containing the vortex axis and the "heavy" axis (here *c*, $m_c > m_{ab}$) of the crystal [5.32]. Obviously this should cause an attraction of vortices within this plane and a formation of vortex chains [5.32, 34]. Vortex separations in such chains depend only on the angle between vortices and the heavy axis so that upon increasing (decreasing) the magnetic field, only the distance between the chains will change.

Experimental studies of the FLL in tilted magnetic fields were performed on single crystals of Bi-Sr-Ca-Cu-O [5.35], Tl-Ba-Ca-Cu-O [5.36] and Y-Ba-Cu-O [5.36]. In these experiments a crystal was tilted with respect to the axis of a solenoid, cooled from $T > T_c$ down to 4.2 K in the external field and then decorated. The angle between the $c$-axis and the holder (which was oriented perpendicularly to the external field) was determined from channeling patterns in the scanning electron microscope, simultaneously with an observation of the decoration patterns [5.36]. The experiments were performed both on the same crystal at different tilt angles and on different samples. In order to find lattice parameters and to measure possible quantitative distortions of the lattice, the images of the vortex patterns were analyzed numerically by taking the two-dimensional Fourier transform. A line through the reflexes on the obtained diffractogram was approximated by an ellipse whose eccentricity corresponded to the $\lambda$ anisotropy; it showed how the FLL deformation is directed.

It was found [5.35, 36] that upon tilting[1] the Bi-Sr-Ca-Cu-O and Tl-Ba-Ca-Cu-O crystals up to 60°, the vortex lattice observed on their surfaces was isotropic (to within an error of $\lesssim 5\%$) and degenerate with respect to the FLL vectors (Fig.5.7). Upon tilting, the area of the FLL unit cell changed approximately as $1/\cos\theta$. Increasing the tilt angle over 60° showed a commensurate array of vortex chains against a background of an isotropic vortex lattice in BSCCO single crystals [5.35]. The chains were uniformly spaced along the sample and oriented in the plane formed by the vortex axes and the heavy axis of the crystal (Fig.5.8). The formation of chains broke the degeneracy of orientational order in the surrounding FLL and aligned one of the close-packed vortex rows parallel to the chains. A study of the effect of magnetic field and tilt angle on the observed patterns revealed that an increase of the magnetic field caused a decrease of both chain separation and vortex separation inside the chains [5.35]. Increasing the tilt angle resulted in an increase of both these separations.

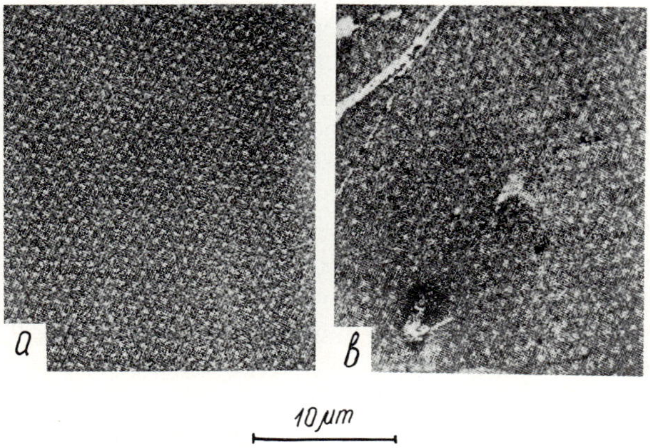


**Fig.5.7.** Isotropic FLL on $Tl_2Ba_2Ca_1Cu_2O_x$ single crystal (**a**); tilted with respect to an external magnetic field, $\theta = 35°$ (**b**)

---

[1] In the experiments the angle between the "heavy" crystal axis and the external field was measured. The vortex direction was assumed to be close to that of the external field since the FLL was "frozen" during the "field cooling" process.

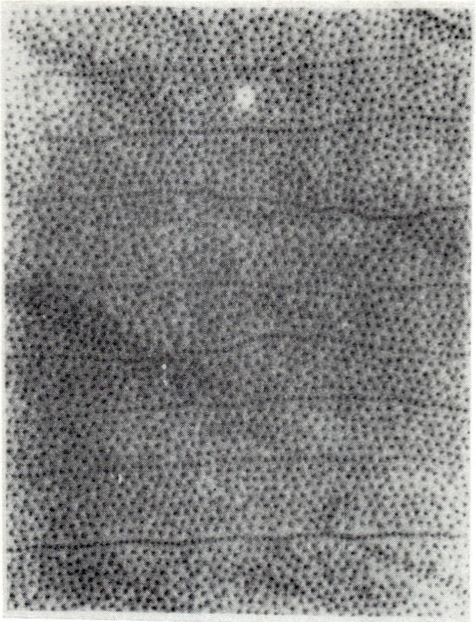

**Fig.5.8.** Illustrationg a region of the decorated crystal at an angle of 70° to an applied magnetic field of 35 G. The dark regions are vortices, with an average spacing of 1.4 $\mu$m. The chains run approximately perpendicular to the rotation axis, and define the orientation of the vortex lattice between chains. (The field of view of 75$\mu$m by 60$\mu$m) [5.35]

The experimentally observed patterns differ from those predicted theoretically, but the most significant result of the theory - the appearance of vortex attraction - holds. The most surprising result of [5.35], the coexistence of an isotropic vortex lattice and vortex chains, is possibly due to a non-equilibrium state of vortex distribution at low temperatures. It may appear in the process of cooling when the sample is going through $H_{c2}(T)$.[2] As recent calculations by *Buzdin* and *Simonov* have shown [5.37], the isotropic FLL is in equilibrium in tilted magnetic fields for $H \gg H_{c1}$. With a decreasing field it becomes metastable and the lowest energy belongs to the vortex chains. However, pinning can lower the mobility of vortices upon cooling hinders the formation of a stable configuration in the FLL so that the isotropic phase can be overcooled. The absence of vortex chains at tilt angles <60° in BSCCO single crystals and up to 80° in TBCCO can be due to similar reasons.

Until now we have not taken into account the energy of the vortex core. In superconductors with a GL parameter $\kappa \gg 1$ the core energy is much smaller than the electromagnetic energy of vortex currents. Nevertheless, it is very important for understanding of pinning processes in high-$T_c$ superconductors. In these materials the coherence length $\xi^{\parallel}$ along the *c*-axis is close to the separation of conducting $CuO_2$ layers, that results in a strongly decreased superconducting order parameter between the layers (and correspondingly the core energy modulation along the *c*-axis). Thus, the energy of the vortex core lying parallel to the layers is considerably smaller than when it is oriented at some angle to them and the vortex can be easily pinned [5.33]. This means that even in the 3D case a vortex may

[2] This interpretation was proposed by A.I. Buzdin

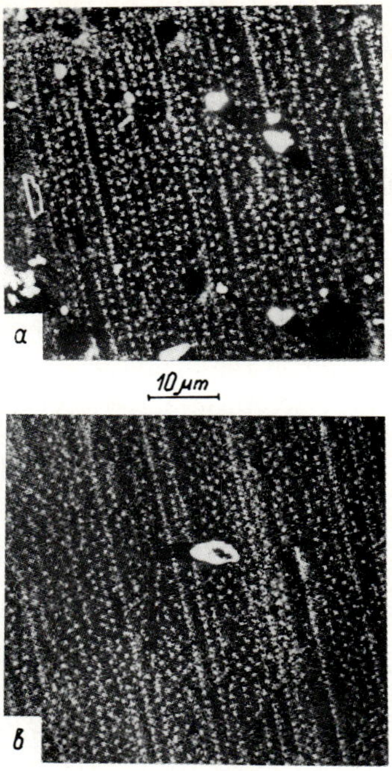


**Fig.5.9.** FLL on identical crystal sections (with similar twin distributions) of $YBa_2Cu_3O_x$ single crystals at different tilt angles and the same value of the external magnetic field $H_e$ = 25 Oe. (**a**) Tilt angle $\theta = 0$, (**b**) $\theta = 35°$

have a stepped structure [5.38]. We believe that such a structure explains the effects observed in our experiments on twinned Y-Ba-Cu-O single crystals (the results are given below).

The aim of the experiments on twinned Y-Ba-Cu-O crystals in tilted magnetic fields was to study the effect of tilting on vortex pinning by twins. As is well known for conventional low-$T_c$ superconductors, effective vortex pinning by plane defects can exist only in a narrow interval of angles between the vortex and the defect plane; for grain boundaries it is $\simeq 5°$ [5.39]. At larger tilt angles the pinning force becomes many times smaller, and for twinned Y-Ba-Cu-O crystals one should expect a change of vortex distribution on twins from that observed for $\theta = 0°$ in [5.40]

Figure 5.9 displays decoration patterns obtained on identical crystal sections (with similar twin structures) at $\theta = 0°$ and $\theta = 35°$ at the same value of the external magnetic field ($H_e$ = 25Oe). One can see that the field tilting did not affect the vortex distribution: the average separation of vortices along the twin boundaries $a_b$ (vortices are attracted to the twin boundaries [5.40]) is practically unchanged as well as the average vortex spacing in monodomain regions $a_v$. As shown in [5.40], the difference $a_v - a_b$ is a measure of the elementary pinning force from the twin boundary. This implies that the pinning force was also not affected by the magnetic-field tilting. On further increase of the tilt angle to $40 \div 45°$, the vortex distribution becomes homogeneous with no difference between twin boundaries and the monodomains region [5.36].

The unusual behaviour is likely to originate from the following. A decrease of the vortex-core energy upon its orientation along a twin boundary (attraction to the boundary) together with the effects of strong anisotropy causes localization of parts of the stepped vortices on twin boundaries. Thus, vertical parts of vortices remain strongly pinned despite the tilting of the magnetic field, and the vortex distribution on the sample surface does not change.

A similar result was obtained from measurements of critical currents in epitaxial thin films of Y-Ba-Cu-O [5.41]. The angle dependence of the critical currents (and, consequently, the pinning force) at temperatures lower than 40 K has a very wide pick (about 40°) when the magnetic field is directed along twin planes. *Tachiki* and *Takahashi* [5.38] used the model of pinned, stepped vortices and obtained expressions which described the $j_c(\theta)$ dependencies to a good accuracy [5.41].

## 5.4 Vortex Pinning by Twin Boundaries

The maximum lossless current in a type-II superconductor - the critical current $J_c$ - is determined by the interaction of Abrikosov vortices and crystal defects (by pinning) which, in turn, causes distortions in the regular FLL. For this reason it is possible to study the current carrying capacity of these superconductors by investigating the distortions resulting from pinning. This proved to be especially helpful in applications to the new class of high-$T_c$ superconductors, since direct measurements of $J_c$ on high-$T_c$ single crystals are, as a rule, unsuccessful because of some problems in the preparation of the samples with proper configuration and contacts to them. Besides, in contrast to the determination of $J_c$ from resistive measurements, as well as from magnetization curves which give integral information of pinning, the direct observation of the FLL deformations enables one to find local pinning forces and, in many cases, to attribute them to particular crystal defects [5.42]. For orthorhombic $YBa_2Cu_3O_x$ single crystals, the most typical defects are twin boundaries. To find the potential of pinning by a twin boundary is an important step in calculating the elementary pinning force $f_p$ and the volume force $F_p = J_cB \sim f_p{}^{\alpha} n^{\beta}$ (B being the magnetic inductance, n is the volume density of defects, $\alpha$ and $\beta$ are the indexes determined by a summation law for elementary forces).

This part of our review, based on [5.40], is devoted to the method of extracting the pinning potential at a single twin boundary from the vortex distribution in the vicinity of the boundary.

### 5.4.1 Calculation of the Pinning Potential

The first direct observation of vortex interaction with twins in $YBa_2Cu_3O_x$ single crystals was reported in [5.14]. An attraction of vortices to the twin boundaries destroying the regular FLL was found. Vortex arrangement adjusts to the twin structure so that the vortices are placed along the twin

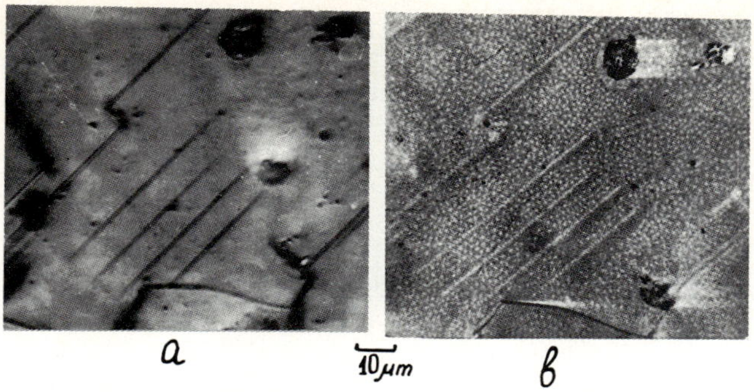


**Fig.5.10.** (**a**) Twin layers (*dark lines*) in the $YBa_2Cu_3O_x$ single crystal. The photo was taken with a microscope using polarized light. (**b**) Vortex distribution in the same section of the crystal (*bright points* are vortices). The photo was taken in a scanning electron microscope. $H_e = 10$ Oe

boundaries with higher density than in the surrounding monodomain regions where the FLL is not destroyed (Fig.5.10). Judging by this, we can assume that the twin boundary represents a potential well for vortices, i.e., a region where superconducting parameters, the lower critical field $H_{c1}$ in particular, differ from those of the bulk. From the difference of vortex separations at the twin boundaries $a_b$ and in monodomain regions $a_v$, one can evaluate the depth of the potential well and, correspondingly, the elementary pinning force per one vortex.

For a superconducting crystal cooled in a magnetic field it is reasonable to expect that the state of the vortex rows at the boundaries and the vortex lattice in the bulk is close to thermodynamic equilibrium. This enables us to evaluate the pinning energy per vortex $U_p$ as follows. At equilibrium, vortex spacings in the volume $a_v$ and at the twin boundaries $a_b$ are determined by the same value of the magnetic field strength $H_i$. (Note that both the field strength $H_i$ and the magnetic inductance B inside the sample differ from the external field $H_e$ [5.13]). In our case (thin platelets in a transverse magnetic field $H_e \ll H_{c1}$) the relations $B \simeq H_e$, $H_i - H_{c1} \ll H_{c1}$ hold). The period $a_v$ in monodomain regions can be determined from a minimization of the vortex lattice energy ($a_v - a_b \gg \lambda$):

$$E_v = n_v \frac{(H_{c1} - H_i)\Phi_0}{4\pi} + \frac{1}{2} n_v \frac{6\Phi_0^2}{8\pi^2\lambda^2} \left(\frac{\pi\lambda}{2a_v}\right)^{1/2} \exp\left(-\frac{a_v}{\lambda}\right), \tag{5.4}$$

where $n_v = 4/(3a_v^2)$ is the vortex concentration, $\Phi_0$ is the flux quantum, and $\lambda$ is the London penetration depth. Minimization of (5.4) with respect to $a_v$ leads to

$$\frac{(H_i - H_{c1})\Phi_0}{4\pi} = \frac{3\Phi_0^2}{16\pi^2\lambda^2} \left(\frac{\pi a_v}{2\lambda}\right)^{1/2} \exp\left(-\frac{a_v}{\lambda}\right). \tag{5.5}$$

The lower critical field at the twin boundary $H_{c1}^*$ is determined by the expression $H_{c1}^* = H_{c1} - 4\pi U_p/\Phi_0$. If the inequality $a_v - a_b >> \lambda$ holds (it is always satisfied in our experiments) then the influence of volume vortices on the vortex rows at the boundaries can be neglected and the period $a_b$ can be determined from a minimization of the vortex-row energy $E_b$

$$E_b = \frac{1}{a_b}\frac{(H_{c1}^* - H_i)\Phi_0}{4\pi} + \frac{1}{2}\frac{1}{a_b}\frac{2\Phi_0^2}{8\pi^2\lambda^2}\left(\frac{\pi\lambda}{2a_b}\right)^{1/2}\exp\left(-\frac{a_b}{\lambda}\right). \tag{5.6}$$

Minimization of (5.6) leads to

$$\frac{(H_i^* - H_{c1})\Phi_0}{4\pi} = \frac{\Phi_0^2}{8\pi^2\lambda^2}\left(\frac{\pi a_b}{2\lambda}\right)^{1/2}\exp\left(-\frac{a_b}{\lambda}\right). \tag{5.7}$$

From (5.5 and 7) we obtain the following expression for the pinning potential $U_p$

$$U_p = \frac{\Phi_0^2}{8\sqrt{2}\pi^{3/2}\lambda^2}\left[\left(\frac{\pi a_b}{2\lambda}\right)^{1/2}\exp\left(-\frac{a_b}{\lambda}\right) - \frac{3}{2}\left(\frac{\pi a_v}{2\lambda}\right)^{1/2}\exp\left(-\frac{a_v}{\lambda}\right)\right]. \tag{5.8}$$

Provided $a_v - a_b >> \lambda$, which is consistent with a large difference between $H_{c1}$ and $H_{c1}^*$, the vortex row parameter $a_b$ must change very slowly. This is valid within the range of magnetic field strengths: $H_{c1} < H_i < H_{c1} + (H_{c1} - H_{c1}^*)$. In the case of a thin plate in a transverse magnetic field, the above interval may correspond to a rather wide range of the external magnetic field $H_e$ due to the large demagnetization factor, so that all the field values in our experiments are included in it.

It is natural to assume that the characteristic distance at which the pinning potential changes is on the order of the coherence length $\xi$. If this assumption is valid then the pinning force per unit length of a single vortex exerted from the twin boundary can be estimated as

$$f_p \simeq U_p/\xi . \tag{5.9}$$

### 5.4.2 Direct Experimental Observation

The decoration experiments were performed in the "field cooling" regime, i.e., after cooling the sample from $T > T_c$ ($T_c$ is the critical temperature) down to 4.2 K with the external magnetic field applied, for several values of the external field $H_e$ from 6.6 to 60 Oe. Both crystals with natural twin structure and untwinned crystals [5.44] which were kindly given to us by M.V. Indenbom and L.A. Dorosinsky were investigated. In order to follow possible changes of vortex structure in the same section of the crystal in different magnetic fields, the sample was cleaned after every decoration procedure and the experiment repeated with the next value of $H_e$. Figure

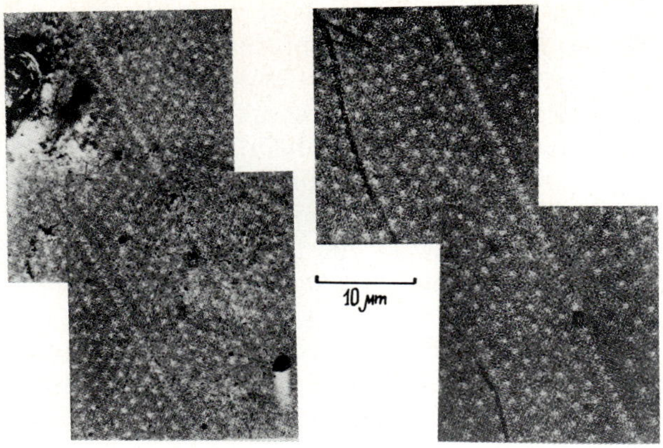


**Fig.5.11.** Vortex distribution in the same section of a crystal but for different magnetic fields. The photos were taken in the scanning electron microscope. (**a**) External magnetic field $H_e$ = 15 Oe; (**b**) $H_e$ = 10 Oe

5.11 shows the distribution of vortices in the same section of the crystal for different magnetic-field values. Upon increasing the magnetic field the average vortex spacing at the twin boundaries $a_b$ decreases to values slower than volume one $a_v$, that is easily seen in Fig.5.12.

For samples with well separated twin boundaries we observed rather small spreads in $a_b$ and $a_v$ values (root-mean-square deviations of $\simeq$10%) both in different samples and in different regions of the same sample. In this case the vortices are uniformly distributed along the twin boundaries. This differs considerably from highly twinned samples where this deviation can reach 100%. These results suggest that the vortex spacings $a_b$ and $a_v$ observed in crystals with low twin density correspond to a stable state and can be used for estimation of the pinning potential at the twin boundary.

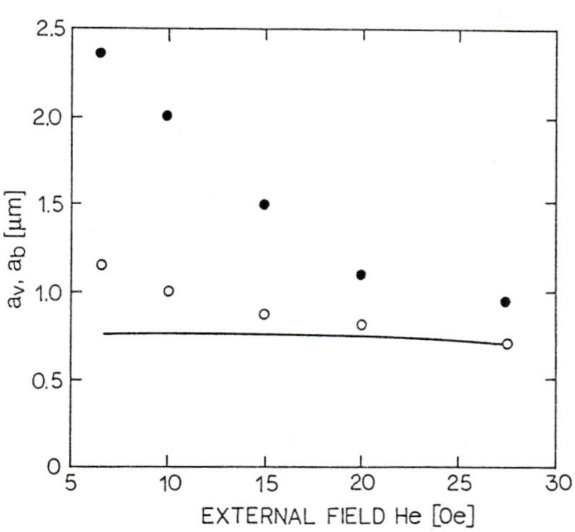


**Fig.5.12.** The dependence of vortex separations on the external magnetic field $H_e$: • - in the volume $a_v$, o - at the twin boundaries $a_b$. The *solid line* shows the function $a_b(H_e)$ calculated from (5.8) for a constant value of $U_p = 3.4 \cdot 10^{-8}$ erg/cm and the experimental values of $a_v$

There is one important aspect in which the above model is different from the experimental situation. High critical currents in Y-Ba-Cu-O crystals with low twin density [5.15] indicate that these crystals contain many other defects, in addition to twin boundaries, most likely point pinning centers. Nevertheless, the greater effectiveness of twin boundaries is beyond doubt. In such a situation vortices have to diffuse in the field of a random pinning potential in order to reach their equilibrium positions. As a consequence, an exchange of vortices between the twin boundary and the bulk can practically be stopped (the structure becomes "frozen") after cooling to some temperature which is, generally speaking, unknown.

Assuming that the vortex distribution observed in the experiment is still consistent with a thermodynamic equilibrium, one can use (5.8) and experimental values for $a_b$ and $a_v$ to calculate the pinning potential $U_p$ which appeared to be increasing with the magnetic field: from $U_p(H_e = 6.6\text{Oe}) = (2.7 \mp 1.3) \cdot 10^{-9}$ erg/cm to $U_p(H_e = 27.5\text{Oe}) = (3.4 \mp 0.8) \cdot 10^{-8}$ erg/cm, i.e., a change of pinning potential by an order of magnitude is obtained. [The value for the penetration depth $\lambda$ was taken as $\lambda(4.2\text{K})$ because the observed vortex structure is believed to have been formed within the temperature region where the difference between $\lambda(T)$ and $\lambda(4.2\text{K})$ is negligible]. However, for a constant difference of the critical parameters in the bulk and at the twin boundary, which is obviously true in our case, there is no cause for the increase of the pinning potential per unit length of one vortex. In addition, for a constant value of $U_p$, the equilibrium parameter $a_b$ must also be unchanged to within the experimental accuracy. [It is easy to show from (5.8) that if $a_v - a_b >> \lambda$ is valid, then the change of $a_b$ must be exponentially small, as demonstrated in Fig.5.12 for $U_p = 3.4 \cdot 10^{-8}$ erg/cm]. At the same time, one can see from Fig.5.12 that when the magnetic field in the sample is increased, the function $a_b(H)$ tends to the one expected for constant $U_p$.

We believe that the above inconsistencies result from the following facts: The vortex structure in very low magnetic fields may not reach the equilibrium state, while the increase of the magnetic field (vortex density) is likely to favor this process due to both shortening of the diffusion path and lowering the barrier for vortices to enter the twin boundary. In other and lowering the barrier for vortices to enter the twin boundary. In other words, it favors a realization of the entire depth of the potential well $U_p$. Proceeding from this, we believe that the value of $U_p$ found with (5.8), for $H_e = 27 \div 5$ Oe, is the closest to the true value of the pinning potential: $U_p \geq 3.4 \cdot 10^{-8}$ erg/cm. The pinning force per unit length of a vortex exerted from the twin boundary is, correspondingly, $f_p \simeq U_p/\xi \geq 0.11$ dyn/cm. When calculating $U_p$ and $f_p$ we used $\xi_{ab}(4.2\text{K}) = 3 \cdot 10^{-7}$ cm and $\lambda(4.2\text{K}) = 1.4 \cdot 10^{-5}$ cm [5.47], corresponding to the temperature used in the experiment. But, indeed, we do not know exactly at what temperature the observed vortex structures were formed. However, according to [5.46], the critical current and, consequently, the pinning forces in the bulk, dropped at temperatures higher than 30 K [$J_c \simeq \exp(-T/T_0)$, where $T_0 \simeq 30$K] and it is reasonable that the observed vortex distribution has become "frozen" at temperatures at least not much higher than 30 K. At the same time at temperatures lower than 70 K the penetration depth $\lambda$ is practically independent of T.

## 5.5 Conclusions

In this section we shall briefly summarize the results obtained from observations of vortex structures with the use of the decoration technique.

A number of novel effects, most of them due to the strong anisotropy of high-$T_c$ superconductors, were found. A detailed investigation of vortex-structure topology gave experimental evidence that not only the Abrikosov vortex lattice can exist in these materials but also new phases, for example, a hexatic glass phase. Convincing data on the anisotropy of Y-Ba-Cu-O crystal and how it is manifested in the vortex structure was the observation of oval vortices on the plane parallel to the *c*-axis and of FLL distortions in the basal plane. These data allow determination of the effective mass anisotropy. Another important consequence of a strongly anisotropic HTSC, namely the appearance of inter-vortex attraction in superconductors with a GL parameter $\kappa >> 1$ when the magnetic field is tilted to the axis of anisotropy, was demonstrated by the formation of vortex chains in decoration patterns. The latter results demonstrate only qualitative agreement with the present theory; therefore further investigations are necessary.

The value of the magnetic flux quantum measured in all the investigated crystals was $2 \cdot 10^{-7}$ G$\cdot$cm$^2$ = hc/2e, the same as in low-temperature superconductors. This is evidence of coupling of carriers in high-$T_c$ superconductors.

Pinning-induced distortions in the vortex structure were found for twinned Y-Ba-Cu-O single crystals. Decoration patterns can show qualitatively whether vortices are attracted to or repelled from a defect. Quantitative estimates of the local individual pinning force are also possible and were made for the pinning force from a twin boundary.

## References


5.1 A.A. Abrikosov: J. Expt. Theor. Phys. **32**, 144 (1957) (in Russian)

5.2 H. Träuble, U. Essmann: Phys. Status Solidi **18**, 813 (1966)

5.3 E.H. Brandt, U. Essmann: Phys. Status Solidi b **144**, 13 (1987)

5.4 R.P. Huebener: *Magnetic Flux Structure in Superconductors*, Springer Ser. Solid-State Sci., Vol.6 (Springer, Berlin, Heidelberg 1979)

5.5 D. Gribier, B. Jacort, M. Rao, B. Farnoux: Phys. Rev. Lett. **9**, 106 (1964)

5.6 H.F. Hess, R.B. Robinson, R.C. Dynes, J.M. Valles Jr., J.V. Waszczak: Phys. Rev. Lett. **63**, 214 (1989)

5.7 U. Essmann, H. Träuble: Phys. Lett. A **54**, 596 (1967)

5.8 L.Ya. Vinnikov, A.O. Golubok: "High Resolution Technique for Direct Observation of the Magnetic Structure at the Surface of Type-II Superconductors"; Preprint, Institute of Solid State Physics, Chernogolovka, USSR (1984)

5.9 A. Ourmazd, J.A. Rentschler, W. Skocpol, D.W. Johnson: Phys. Rev. **36**, 8914 (1987)
5.10 G.A. Emelchenko, M.V. Kartsovnik, P.A. Kononovich, V.A. Larkin, Yu.A. Ossipyan, V.V. Ryazanov, I.F. Schegolev: J. Expt. Theor. Phys. Lett. **46**, 162 (1987) (in Russian)
5.11 F.L. Gammel, D.J. Bishop, C.J. Dolan, J.R. Kwo, C.A. Murray, L.F. Schneemeyer, J.V. Waszczak: Phys. Rev. Lett. **59**, 2592 (1987)
5.12 L.Ya. Vinnikov, L.A. Gurevich, G.A. Emelchenko, Yu.A. Ossipyan: JETP Lett. **47**, 131 (1988)
5.13 L.Ya. Vinnikov, L.A. Gurevich, G.A. Emelchenko, G.A. Kazaryan, N.N. Kolesnikov, M.P. Kulakov, D.Ya. Lenchinenko, Yu.A. Ossipyan: Solid State Commun. **70**, 1145 (1989)
5.14 L.Ya. Vinnikov, L.A. Gurevich, G.A. Emelchenko, Yu.A. Ossipyan: Solid State Commun. **67**, 421 (1988)
5.15 G.D. Dolan, G.V. Chandrashekhar, T.R. Dinger, C. Field, F. Holtzberg: Phys. Rev. Lett. **62**, 827 (1989)
5.16 R.N. Kleiman, P.L. Gammel, L.F. Schneemeyer, J.V. Waszczak, D.J. Bishop: Phys. Rev. Lett. **62**, 2331 (1989)
5.17 E.M. Chudnovsky: Phys. Rev. B **40**, 11357 (1989)
5.18 M.C. Marchetti, D.R. Nelson: Phys. Rev. B **41**, 1910 (1990)
5.19 C.A. Murray, P.L. Gammel, D.J. Bishop, D.B. Mitzi, A. Kapitulnik: Phys. Rev. Lett. **64**, 2312 (1990)
5.20 W.E. Lawrence, S. Doniach: *Low Temperature Physics*, Proc. 12th Int'l Conf., Kyoto, 1970 (Keigaku, Tokyo 1971) p.361
5.21 R.A. Klem, A. Luther, M.R. Beasley: Phys. Rev. B **12**, 877 (1975)
5.22 M. Tuominen, A.M. Goldman, Y.Z. Chang, P.Z. Jiang: Phys. Rev. B **42**, 412 (1990)
5.23 S.N. Artemenko, I.G. Gorlova, Yu.I. Latyshev: Phys. Lett. A **138**, 428 (1989)
5.24 L.N. Bulaevskii: J. Expt. Theor. Phys. **37**, 1133 (1973)
5.25 V.L. Ginzburg: J. Expt. Theor. Phys. **23**, 236 (1952) (in Russian)
5.26 A.M. Grishin, A.Yu. Martynovich, S.V. Yampol'skii: "Magnetic Flux in Anisotropic London-Type Superconductors"; Preprint, Donetsk Physicotechnical Institute, Donetsk USSR (1988)
5.27 L.Ya. Vinnikov, I.V. Grivor'eva, L.A. Gurevich, Yu.A. Ossipyan: J. Expt. Theor. Phys. Lett. **49**, 99 (1989)
5.28 G.J. Dolan, F. Holtzberg, C. Feild, T.R. Dinger: Phys. Rev. Lett. **62**, 2184 (1989)
5.29 F. Izumi, H. Asano, T. Ishigaki, A. Ono, F.P. Okamura: Jpn. J. Appl. Phys. **26**, L611 (1987)
5.30 V.K. Vlasko-Vlasov, M.V. Indenbom, Yu.A. Ossipyan: J. Expt. Theor. Phys. Lett. **47**, 375 (1988)
5.31 W. Bauhofer, W. Biberacher, B. Gegenheimer, W. Joss, R.K. Kremer, Hj. Mattausch, A. Müller, A. Simon: Phys. Rev. Lett. **63**, 2520 (1989)
5.32 A.M. Grishin, A.Yu. Martynovich, S.V. Yampol'skii: J. Expt. Theor. Phys. **97**, 1930 (1990) (in Russian)
5.33 D. Feinberg, C. Villard: Phys. Rev. Lett. **65**, 919 (1990)
5.34 A.I. Buzdin, A.Yu. Simonov: J. Expt. Theor. Phys. Lett. **51**, 168 (1990) (in Russian)
5.35 C.A. Bolle, P.L. Gammel, D.G. Grier, C.A. Murray, D.J. Bishop, D.B. Mitzi, A. Kapitulnik: Phys. Rev. Lett. **66**, 112 (1990)
5.36 L.A. Gurevich, I.V. Grigorieva, N.N. Kolesnikov, M.P. Kulakov, V.A. Larkin, L.Ya. Vinnikov: Physica C **195**, 323 (1992)
I.V. Grigorieva, L.A. Gurevich, L.Ya. Vinnikov: Physica C **195**, 327 (1992)

5.37 Y.I. Buzdin, A.Yu. Simonov: private communication
5.38 M. Tachiki, S. Takahashi: Solid State Commun. 2, 1083 (1989)
5.39 W.E. Yetter, E.J. Kramer: J. Mater. Sci. **17**, 2792 (1982)
L.Ya. Vinnikov, V.G. Glebovskii, S.I. Moscvin: J. Expt. Theor. Phys. Lett. **33**, 253 (1981) (in Russian)
5.40 L.Ya. Vinnikov, I.V. Grigor'eva, L.A. Gurevich, A.E. Koshelev: Superconductivity. Phys. Chem. Techn. **3**, 1385 (1990)
5.41 B. Roas, L. Shultz, G. Saemann-Ischenko: Phys. Rev. Lett. **64**, 479 (1990)
5.42 H. Träuble, U. Essmann: Phys. Stat. Solidi **32**, 337 (1969)
C.P. Herring: J. Phys. F: Metal Phys. **6**, 99 (1976)
I.V. Grigoryeva, L.Ya. Vinnikov: J. Low-Temp. Phys. **74**, 81 (1989)
5.43 L.D. Landau, E.M. Lifshits: *Electrodynamics of Continuous Media* (Pergamon, Oxford 1984)
5.44 L. Dorosinsky, B. Farber, M. Indenbom, V. Nikitenko, A. Polyanskii, V. Vlasko-Vlasov: Ferroelectrics **111**, 321 (1990)
5.45 P.H. Kes, J. van den Berg: Flux pinning and thermally activated depinning in single crystals of high-temperature superconductors, in *Studies of High-Temperature Superconductors*, ed. by Narlikar, Anant (Nova Science, New York 1989)
5.46 M. Oda, Y. Hidaka, M. Suzuki, T. Murakami: Phys. Rev. B **38**, 252 (1988)

# 6. Magnetization Processes

V.K. Vlasko-Vlasov, M.V. Indenbom, and A.A. Polyanskii

Despite the huge number of publications on the magnetic properties of high-$T_c$ superconductors many principal questions concerning magnitude and anisotropy of critical fields and currents, the unusual reversibility of magnetization, the role of defects on the superconductivity, the extremely short coherence length, etc. are still unanswered.

The most detailed and objective information about the magnetization features of HTSC samples can be obtained from direct observations of the magnetic-flux behaviour in the samples under different experimental conditions. We shall describe a new technique for the magnetooptic visualization in a wide temperature range of the flux distributions in superconductors based on iron-garnet films. Results for HTSC single crystals, films, and ceramics are presented. Features of the mixed state, penetration field, temperatures dependences and the anisotropy of the critical currents, and effects of twins have been investigated and are discussed.

## 6.1 Magnetic Studies of High-$T_c$ Superconductors

Studying the behavior of a material in a magnetic field comprises the most essential test of its superconducting properties. In the case of type-II superconductors the fundamental magnetic characteristics, in addition to the Meissner effect, are the lower and upper critical fields, determined respectively by the magnetic penetration $\lambda$ and the coherence length $\xi$, and also the critical current, associated with the flux-line pinning. Measurements of their values, of the anisotropy and the temperature dependences give important information about the nature of superconductivity.

Despite the continuing discussion about the coupling mechanism, it has reliably been established by now that new high-$T_c$ compositions are typical type-II superconductors, possessing a high Ginzburg-Landau constant $\kappa = \lambda/\xi$ (reviews by *Malozemoff* [6.1] and *Meilikhov* [6.2]). However, the strong anisotropy of their physical parameters, stipulated by the complicated perovskite structure (for example, in the least anisotropic case of Y-Ba-Cu-O at T = 0 K, $\xi_{ab} \sim 16$ Å, $\xi_c \sim 3$ Å [6.3], $\lambda_c \sim 4500$ Å, $\lambda_{ab} \sim 1300$ Å [6.4], where the saubscripts indicate the crystal axes ***a***, ***b*** and ***c***), low pinning barriers (10÷100 meV [6.5]). They are strongly dependent on the temperature T and field H. The high transition temperatures which resulted in a series of peculiar features in their magnetic properties have not been observed in conventional superconductors. These are a strong dependence of the Meissner fraction on the field value and a vanishing of the difference between magnetizations obtained under different cooling conditions (field cooling and zero-field cooling) at certain fields $H^*(T)$ (in traditional superconductors the difference is observed up to $T_c$); a considerable broadening of the resistive transition in the field; a large relaxation of the magnetic flux dependent logarithmically on the time (but not exponentially as in low-$T_c$ materials); a strong decay of the critical current against T and H [6.1,2]. These features required the definition of new concepts in superconductivity terminology such as an irreversibility line, a giant flux creep, a vortex glass state, and so on.

Transitions in the vortex structure are actively discussed in connection with a dimensionality crossover, when 3D anisotropic flux lines turn into chains of weakly coupled 2D vortex "beads", localized in the superconducting $CuO_2$ planes, (see, e.g., [6.6]). This problem which was brought up earlier in connection with low-$T_c$ layered superconductors by *Lowrence* and *Doniach* [6.7], and *Bulaevskii* [6.8], has become highly topical for describing the anomalous temperature and field dependences of the critical current $J_c$ in HTSC [6.1,6,9].

It is worth noting that the critical-current problem associated with ascertaining effective mechanisms and with the role played by different pinning centers is the key problem for HTSCs. This issue is exactly the one which determines the possibilities for technical application of these materials.

Defects influence the HTSCs' physical properties. Because of the short coherence length, defects of even an atomic scale must be considered as essential pinning centers [6.10]. They can also result from weak links and from a Josephson junction network side by side with interlayers between superconducting $CuO_2$ planes.

In HTSC ceramics the dominating effects due to weak links at the grain boundaries, determining low-$J_c$ values, are quite clear [6.11]. For this case the superconducting glass model by *Müller* et al. [6.12] seems quite natural. It explains the experimentally observed irreversibility line $H^* \sim (1-T)^{3/2}$, analogous to the Almeide-Touless line in spin glasses, and describes the strong logarithmic magnetic relaxation as a direct consequence of the destruction of randomly arranged weak links.

It turned out, however, that high-$T_c$ single crystals possess similar properties, too. In addition to the irreversibility line they exhibit nonmonotonic dependences for M(H) and $J_c$, considered as possible evidence of their granular structure [6.13, 14].

A possible origin for the weak-link formation in crystals are the twin boundaries, as suggested by *Deutscher* and *Müller* [6.15]. However, there are also calculations predicting enhanced superconductivity at the plane defects. In particular, *Khlyustikov* and *Buzdin* demonstrated that a $T_c$ increase should be expected at the twin walls [6.16]. *Abrikosov* et al. calculated the latter effect for HTSC [6.17] and suggested an explanation for the behavior of the temperature change of the upper critical field near $T_c$ [6.18] and for two anomalies of the specific heat in the vicinity of the superconducting transition [6.19,20]. The problem of twin-boundary effects still remains open because the researchers advancing both opposing views find support in experimental results.

At high temperatures the processes of relaxation of the vortex become the most important ones in determining the subsequent behavior of a sample. Their interpretation in terms of flux creep and flow, for the case of HTSC, is a serious alternative to the superconducting glass model and seems more reasonable in many cases. In fact, the irreversibility line and the logarithmic magnetization decay were well treated by *Yeshurun* and *Malozemoff* [6.21] on the basis of the traditional creep model, proposed by *Anderson* and *Kim* [6.22,23], taking into account HTSC features.

To understand the flux behavior in HTSCs the dimensionality change mentioned above and new features of the phase diagram of the vortex structure found recently by theorists, such as the formation of a vortex glass phase [6.24,25] and various regions of existence of a vortex liquid [6.26], are of importance.

Taking into account the thermoactivated flux creep, the difference between transport and magnetic critical current values can be explained [6.27]. Effects of the creep also manifest themselves in experimental values and in temperature changes of apparently static parameters such as the upper $H_{c2}$ [6.28] and low $H_{c1}$ [6.29] critical fields.

One should note that there is a considerable spread in experimental data for $J_c$, $H_{c1}$ and $H_{c2}$, which is associated not only with different contributions of the creep observed under different measuring conditions, but, first of all, with the high structure sensitivity of the superconducting characteristics. This is due to the large defect-structure mobility in HTSC crystals, which can be changed by even the slightest variations of the growth conditions and thermal treatment of samples. This is determined, to a considerable extent, by the fast diffusion of oxygen [6.30], the concentration and ordering of which strongly change the transition temperature and other characteristics [6.31,32]. At present a technique for reproducibly manufacturing homogeneous HTSC samples with stable properties has still not been developed, although there are ways for improving them (e.g., low-temperature oxygen annealing). Therefore, the problem of understanding the effects of structure inhomogeneities on the superconductivity of new materials becomes more important.

### 6.1.1 Experimental Techniques

To investigate the magnetic properties of HTSCs all the known methods developed for conventional superconductors are used. These are, first of all, measurements of the temperature and field dependences of magnetization M(H,T) (carried out in most cases using the SQUID or vibration magnetometers). They give the critical field values $H_{c1}$ and $H_{c2}$, and determine the critical current $J_c(H,T)$ from the difference of M values obtained at increasing and decreasing fields. The transport data also provide $J_c$ measurements. In addition, from the resistance curves $\rho(H,T)$ the $H_{c2}$ value is derived (with a correction due to creep), and changes of the vortex structure states (the vortex lattice melting, depinning and so on) are revealed. These changes are also determined using the vibration techniques, which detects the mechanical losses in superconductors [6.33, 34]. Pinning effects and the anisotropy are studied by mechanical torque measurements at varying field directions [6.35]. Information about the magnetic penetration depth (although only for samples of large mass) is obtained from the $\mu$SR experiments [6.36, 37].

The above methods are macroscopic in nature and do not allow studying specific effects from inhomogeneities of a sample. Only some qualitative anisotropic features due to defects can be revealed. (E.g., from orientational anomalies of M(H, T) and torque measurements some correlation between the average pinning value and the vortex-line direction with respect to twin boundaries can be observed [6.38-40]). To obtain exact data about the details of the mixed state in HTSCs, about the role of defects in the vortex penetraion and trapping processes, about the pinning anisotropy and so on, direct methods for magnetic flux observations are required. The most useful tools are the decoration and magnetooptic techniques. The decoration method was first applied by *Balashova* and *Sharvin* [6.41] to observe normal phase domains in type-I superconductors. Then, *Essmann* and *Träuble* [6.42] developed this method for type-II superconductors. In HTSCs the first direct evidence of the carrier pairing below $T_c$ was obtained by *Gammel* et al. [6.43] and *Vinnikov* [6.44], and an effect of twins on the magnetic-field penetration was revealed [6.45, 46] using the decoration techniques. Unfortunately, these experiments can be conducted at helium temperatures and under field-cooling conditions only. Powder patterns obtained at higher temperatures have no simple interpretation [6.47]. It is worth noting that single vortices can be also observed by a tunneling microscope [6.48], however, until now such an experiment has not been done in HTSCs.

Not only magnetic flux observations but also studies of the kinetics of flux changes are possible with the magnetooptic methods based on the optical visualization of the field at the surface of a superconductor. In conventional materials EuS-F layers deposited directly on the samples were successfully used for analyzing the flux topology and intensity [6.49]. Their applicability to HTSCs was recently demonstrated by *Moser* et al. [6.50], but their working temperature range is at present restricted by the upper limit of 10 K.

*Batalla* et al. [6.51] showed the possibility of flux observation in Y-Ba-Cu-O crystal at T = 4.2 K using a magnetooptic lead glass. However, to provide sufficient magnetooptic quality the glass needed to be thick enough (~0.2 mm) which limited the resolution of the method.

Recently, we have succeeded in adapting ferrimagnetic-garnet thin films for visualizing magnetic flux in HTSCs. These films have a large Faraday rotation and can be employed in a wide range of temperatures and fields [6.52]. This method was subsequently used by other groups [6.53, 54]. A direct study of the magnetic-field penetration and trapping processes in single crystals, films and ceramic samples of HTSCs was conducted with these iron-garnet films. The basic goal was to analyze the flux structure and to reveal its relation to sample imperfections [6.52, 55-59]. The main results are presented in the following sections. Special attention was paid to the weak superconducting regions (granularity) in single crystals and films, to the effects of twins on the superconducting parameters, to peculiarities of the flux structure for different field directions and, accordingly, to the anisotropy in the magnetic properties.

In practically all the studied crystals (of different structures and compositions) stages of the reversible and then, with growing field, irreversible flux penetration were found. They are associated with the presence of areas with higher and lower superconducting characteristics and with bending of flux lines determined by the anisotropy and the sample shape. It is established that twins have an insignificant influence on $T_c$ compared with other crystal imperfections. Their effect is revealed only in the magnetic-flux pinning at sufficiently high temperatures. Measured values of the critical current, and their temperature changes and the anisotropy are correlated with those reported by other researchers. The advantages of the method for the analysis of a HTSC sample are quality evaluated.

## 6.2 Magnetic-Flux Visualization and Measurement of Local Parameters

In the present work commercial thin ($\lesssim 10\,\mu$m) films of ferrimagnetic garnets doped by bismuth (increasing the Verdet constant) on Gd-Ga garnet substrates were used.[1] The initial magnetic-domain structure of these films is a labyrinth of domains, magnetized normal to the film plane. Their width (less than the film thickness) is essentially the limit of the method's resolution.

Domains can be seen in a polarizing microscopy by their different colorings when polarizers are uncrossed or they are seen as bands of equal intensity with dark boundaries when the polarizers are crossed. When the field **H** is normal to the film-surface domains magnetized along **H** expand and their neighbors contract. The width changes can easily be assigned to a field value.

In the experiment the film is placed on the flat sample surface of an as-grown crystal, film or polished ceramics, and the domain structure is analyzed in the reflected light passing twice (down and up). The pattern's contrast is increased if the lower surface of the film is sputtered with Al to improve the reflection coefficient. However, a satisfactory picture can be obtained even without sputtering. In the latter case it is possible to observe a superconducting sample surface whose image is superimposed on the do-

[1] Magnetooptic indicator garnet films were grown by the scientific industrial association "Gamma" (Zelenograd, Russia).

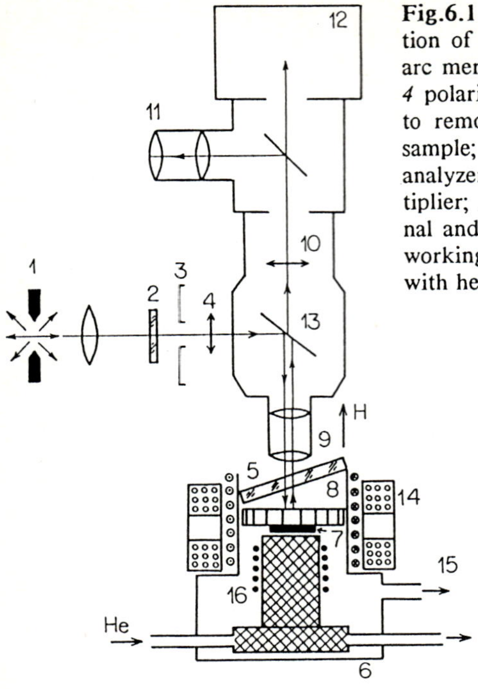


**Fig.6.1.** Experimental setup for the observation of magnetic fluxes in superconductors. (*1* arc mercury lamp; *2* color filter; *3* diaphragm; *4* polarizer; *5* window of the cryostat inclined to remove parasitic reflections; *6* cold bar; *7* sample; *8* indicator film; *9* objective lens; *10* analyzer; *11* eyepiece; *12* camera or photomultiplier; *13* beam splitter; *14* coils for longitudinal and transverse fields; *15* evacuation of the working volume of the cryostat; *16* cold finger with heater)

main pattern in the indicator film. This shows the direct relationship between features of the magnetic-flux structure and sample imperfections. In particular, the effect of twin-structure inhomogeneities in 1-2-3 crystals, easily revealed in the reflected light, can be observed [6.60-62].

The experimental setup is shown in Fig.6.1. Samples are placed in an optical micro-cryostat mounted at the microscope stage. A specially designed cold finger equipped with a heater allowed the temperature to change from 6 K to ~300 K and back again within several minutes. The temperature is measured by a Cu-Fe thermocouple caulked into the copper substrate directly under the sample.

When a sample is in the nonsuperconducting state, field-induced changes of the domain structure are homogeneous across the indicator field area. However, below $T_c$ the film is magnetized around the sample and above it the field is screened so that the domain structure in the films remains unchanged. An example of the Meissner screening can be seen in Fig.6.2a. In the case of flux trapping the film is magnetized above those sample regions containing the flux, and around them unmagnetized domains are observed in the film. When the flux is pinned over the entire sample, there is a band around it remagnetized in the opposite direction by stray fields of the trapped flux (Fig.6.2c). Note that the domain width in the film near the edge of a sample in the Meissner state gives an estimate of the field concentration, i.e., the demagnetizing factor N of the sample $H_{edge} = H/(1-N)$.

The above method was developed for measurements of the local magnetic susceptibility curves at different sample points [6.56]. For this purpose, in the image field of the microscopy a region was selected by using a diaphragm of diameter exceeding the width of the magnetic-domain image, and the amplitude A of the domain-wall vibrations in this region under an AC external field was measured. The value of A (detected from optic measurements using a photomultiplier and a lock-in amplifier) is determined by the field amplitude in the chosen film area, and below $T_c$ above the sample it will go to zero due to superconductive screening. Thus the temperature dependences of A in different points directly correspond to the susceptibility curves but, unlike the macroscopic $\chi(T)$ changes, they reveal local inhomogeneities of $T_c$ across a sample. If regions of stronger superconductivity (higher $T_c$) were arranged continuously along the sample perimeter this technique would not allow measurement of $T_c$ variations inside such a sample. But, in fact, these variations were detected in all crystals studied by the above modulation method.

To determine the critical current, the Shubnikov-phase penetration depth X for $H > H_{c1}$ was measured as a function of the field strength. Taking into account that induction at the sample edge is equal to the external field corrected by the demagnetizing factor, the critical current can be estimated from the relation $4\pi J_c/c = H/X(1-N)$ [the applicability of the formula was justified by the observation of the linearity of X(H)]. The necessity of improving this approximation taking into account the dependence on the field direction (associated with the flux line bending) will be discussed below.

## 6.3 Experimental Results and Discussion

### 6.3.1 Low-Field Magnetization in High-$T_c$ Single Crystals. Magnetic-Flux Penetration and Trapping at H||*c*

First, HTSC-crystal magnetization, which is strongly affected by demagnetizing fields when the external field **H** is parallel to the ***c***-axis, will be described. The samples were plates with a wide basal plane ***ab*** and a smooth as-grown surface. Some of the yttrium ($YBa_2Cu_3O_{7-\delta}$) and thallium ($Tl_2Ba_2CaCu_2O_x$) cuprate crystals had rectangular edges and some were of irregular shape. They had in-plane dimensions from several hundreds of micrometers up to millimeters and their thickness ranged from 20 to 200 $\mu$m. $Bi_2Ca_2SrCu_2O_x$ crystals were in the form of thin plates (up to 30$\mu$m) with irregular edges.

Common features of the magnetization processes for different chemical compositions under fields of up to ~1 kOe will be illustrated by an example of the Y-Ba-Cu-O. Then some features observed in Bi [6.55] and Tl [6.56] samples will be mentioned.

When **H**||***c*** the indicator film is placed on the wide crystal surface and the field is applied normal to it. For $T < T_c$ the film first magnetizes out-

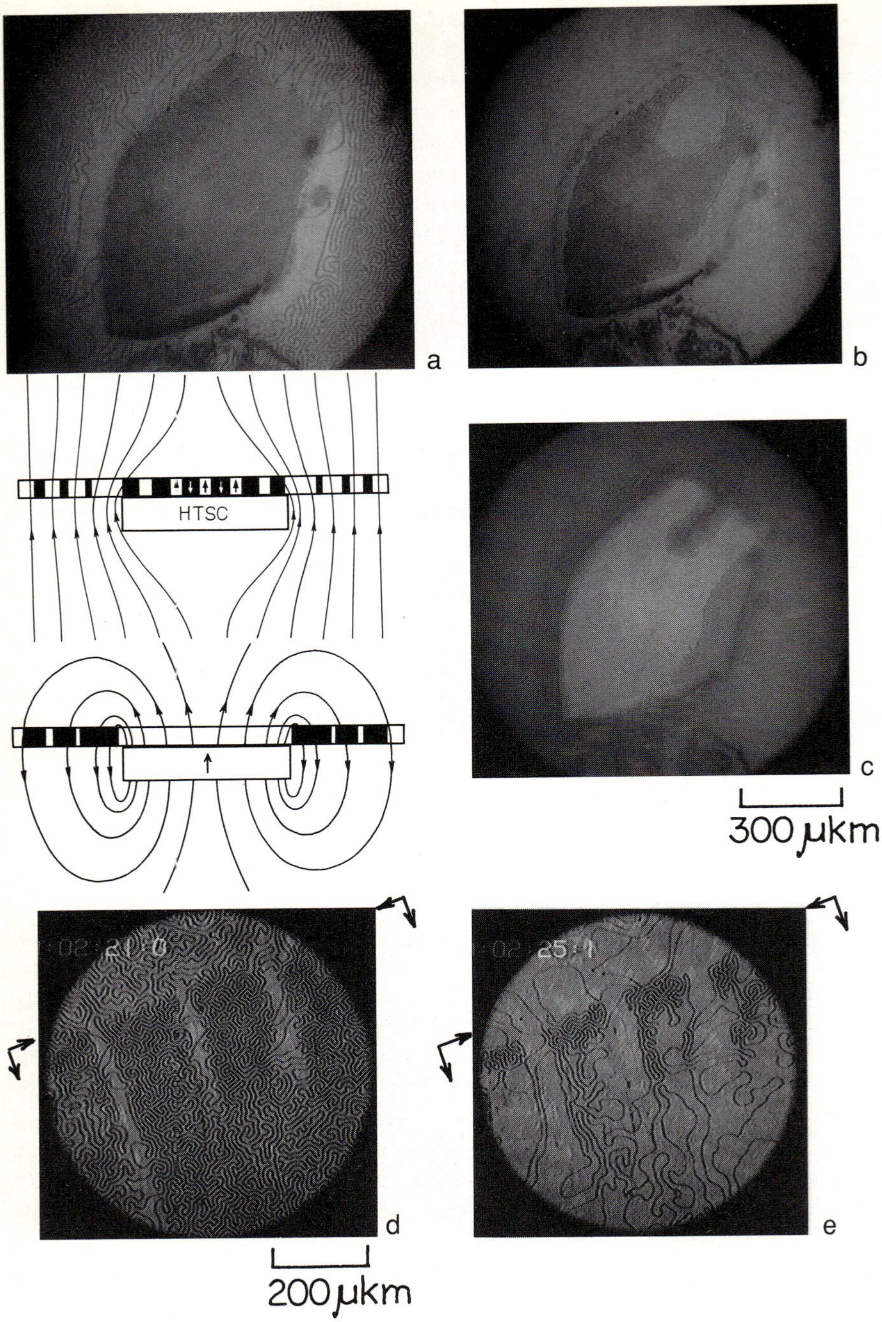


side the sample with increasing field, which is stronger due to the demagnetizing effect at the crystal edges (as in Fig.6.2). When some critical value $H_1$ is reached, the flux begins to penetrate from the sample edges, and the film begins to magnetize in this area. A magnetization front moves from the

Fig.6.2a-e. Magnetic field penetration and trapping patterns observed above HTSC samples using garnet films. (**a**) Screening of the field **H**||*c* by a Y-Ba-Cu-O single crystal (scheme shown below); the field is concentrated near the sample edges (H = 80Oe; T = 20K); (**b**) inhomogeneous field penetration: a region of easy penetration (weak region) is revealed at the top of the sample (H = 390Oe, T = 20K); (**c**) trapped flux (scheme shown at the left) observed after cooling in a field of 1125 Oe (the field is switched off, T = 20K), the weak region is remagnetized in the opposite direction by the demagnetizing field of the trapped flux; (**d**, **e**) penetration of the field in a Y-Ba-Cu-O film (T = 20K, H = 100Oe and 180Oe, respectively), strips of easy field penetration and regions of stronger screening near the film edge (film position is indicated by *arrows*) are seen

edges to the crystal center with increasing H; and the magnetization pattern depends strongly on the temperature. At low temperatures the penetration front in most cases follows the crystal shape (see insert of Fig.6.4). However, with increasing T practically in all studied samples an inhomogeneous field penetration was observed. The flux progressed to different distances from the crystal edges, thus revealing regions of easier field penetration (Fig.6.2b). Under these conditions, areas of strong screening where the flux enters last are often found to be not in the crystal center but at the edges where the field concentration is known to be maximum.

The penetration picture unexpectedly turned out to be reversible up to field strengths considerably exceeding the $H_1$ value. That is for decreasing field values (even when the magnetization front reached the crystal center) the flux began to escape, and its boundary reversible (with an insignificant hysteresis) moved from the center towards the sample periphery. This phenomenon was not observed but in HTSC films, where, in accordance with classical notions, the flux escapes with decreasing H at the sample edges, forming a band free of vortices.

Only after reaching some field $H_2 > H_1$ does the trapped flux remain in the crystals. Both the penetration field $H_1$ and the trapping field $H_2$ increase with decreasing temperature. As a rule, $H_1(T)$ is linear in agreement with data of many researchers for the lower critical field $H_{c1}(T)$ in HTSC crystals [6.63-67], and the slope of $H_2(T)$ increases with decreasing T (Fig.6.3). Note that some magnetic data, considered as measurements of the first critical field [6.29,67], give a similar $H_{c1}(T)$ slope increase at low temperatures.

Generally speaking, values of $H_1$ and the slopes of their temperature dependence are different for regions where the penetration is easy or hard. The presence of such regions implies that there are weak links resulting in a variation of superconducting parameters across a sample. These variations were observed by using the modulation method. As expected, the local susceptibility curves measured in regions of easier field penetration show $T_c$ below that obtained in regions of strong screening. Macroscopic curves of $\chi(T)$ exhibit steps at the corresponding temperatures. The $T_c$ variations will be discussed in Sect.6.3.5 in more detail.

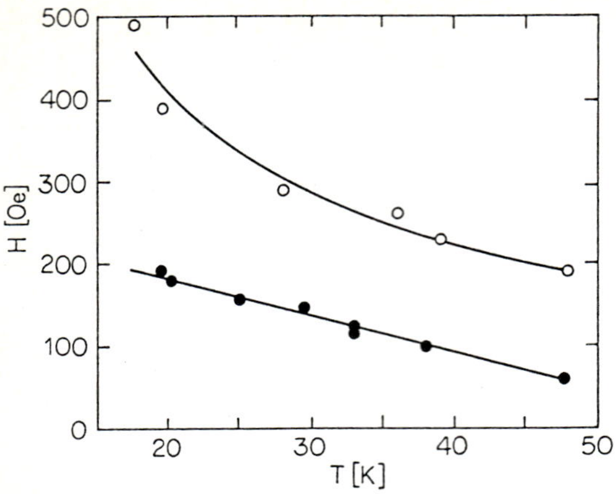


**Fig.6.3.** Temperature dependence of the penetration (•) and trapping (o) fields in a Y-Ba-Cu-O single crystal

Areas with different properties are also revealed from the flux trapping patterns obtained under field-cooling conditions. Regions with higher $T_c$ values, that showed stronger screening at zero-field cooling, were generally better at trapping the flux over a wide temperature range. But in areas of easy field penetration, flux escape was also easier. Such behavior was observed both under field cooling and after application of a sufficiently high field. However, in the case of field cooling, the flux was already trapped at field values at which the magnetization is reversible in a zero-field cooling experiments. This corresponds to the difference in the Meissner fractions for the two methods of cooling known from macroscopic M(H) measurements [6.1]. It should be noted that at low temperatures (<20K) sometimes the strongest residual flux was observed in the regions of easy field penetration, as if they were the regions of most effective pinning. This can be explained by the presence of areas with higher superconducting properties around the weak regions acting as a barrier to vortex escape.

These facts suggest that the reversibility of flux penetration at fields below $H_2$ is associated with the granularity of HTSC crystals [6.57]. Formation of weak regions (granularity) in HTSC single crystals was noted by many researchers. *Sulpice* et al. [6.68], from M(H) curves measured for crystals of different sizes, concluded the presence of weak links inside them which limit the spatial scale of superconducting currents. *Daemling* et al. [6.13] treated the nonmonotonic M(H) curve and the broadening of the superconducting transition under a magnetic field as results of the granularity effect. Both groups believe that the weak regions are produced by depleted oxygen content in the crystals. *Hibbs* and *Campbell* [6.69], and *Küpfer* et al. [6.14, 70] observed a sharp drop in the field dependence of the critical current (obtained from magnetic measurements) at low H values. Then, $J_c$ increases and only at stronger fields does it begin to decrease again. The first region of the $J_c$ decay is associated with weak regions that are formed, as the above-mentioned researchers believe, at the twin boundaries. The penetration of very low fields in Y-Ba-Cu-O single crystals, which was detected by microwave absorption experiments by *Dulčic* et al. [6.71], is also ascribed to weak links formed at the boundaries and other plane defects. According to the classical critical-state model of *Bean* [6.72] the flux should penetrate into a superconductor for $H > H_{c1}$ and the flux density should decay inside it with a gradient determined by the critical current (or the bulk pinning force). After switching off the field, vortices should escape along the sample boundary, here forming a flux gradient of the same value but directed to the outside. In [6.57] it was supposed that for $H_1 < H < H_2$ the flux enters a crystal along weak link channels. The channels can be considered to have a random length distribution; however, their density is inhomogeneous - it is higher in the regions of weak screening. Then there are long channels starting at the sample edges and leading from weak to stronger superconducting areas. In addition, there are more numerous, but shorter, channels running from the edge in the same direction. As a result the flux penetrating along the channels has a gradient directed to the crystal interior from the weak to strong areas. Low pinning in the channels can explain the reverse flux escape with decreasing field.

In [6.45, 46] the vortex pattern observed using the powder technique in field-cooled Y-Ba-Cu-O crystals shows vortex chains concentrated along the twin boundaries. The possibility is not excluded that this is associated with the formation of easy penetration channels along the twin boundaries where, according to [6.15], a weakening of superconductivity could take place. It will be shown in Sect.6.3.5 that a stronger effect on $T_c$, compared to twins, exerted by the oxygen concentration variations in HTSC crystals. Incidentally, the oxygen content can be reduced just at the twin boundaries [6.73].

Note that the existence of weak areas can result in decreasing the demagnetizing factor observed in some samples at rising temperature [6.74]. Obviously, at low T the total sample volume screens the field (or the supercurrent path follows the crystal shape), but after heating it is divided into weakly coupled regions so that the total demagnetizing factor drops.

Thus penetration of the field along the weak channels can be reversible. However, with increasing H the flux begins entering the surrounding regions of stronger superconductivity around weak channels and is essentially trapped there. This stage is characterized by the field $H_2$. Perhaps it can be considered as the lower critical field for strong regions. In [6.67], increasing of the slope of the $H_{c1}(T)$ at low temperatures, similar to our observation of $H_2(T)$ in Fig.6.3, is explained by the contribution of a lower-$T_c$ phase with higher $H_{c1}(0)$. But in [6.29] such temperature changes are ascribed to an increase of the penetration field, determined by the creep of vortices through the Bean-Livingston barrier, and not to $H_{c1}(T)$ variations.

Linear changes of $H_1(T)$ seen in Fig.6.3 can be compared with the $H_{c1}(T)$ dependence predicted by the Ginzburg-Landau theory, where $\lambda \sim (1-T/T_c)^{-1/2}$ and $H_{c1} \simeq \phi_0/4\pi\lambda^2 \sim (1-T/T_c)$. Certainly one must keep in mind that this theory is applicable at not very low temperatures. Below, it will be shown that the penetration field in the case $\mathbf{H}\|c$ can give only an underestimated value of $H_{c1}$ due to vortex bending (which assists the reversibility and changes the screening current components).

It is worth noting that it is possible to explain the magnetization reversibility in fields slightly exceeding $H_1$ in terms of the vortex-structure phase diagram [6.26]. In the mean-field approximation the $H_{c1}(T)$ curve is known to separate the Meissner state of a superconductor from the phase of the vortex lattice. The period of the latter is reduced with increasing H until vortex cores begin to overlap at the phase boundary $H_{c2}(T)$ where a superconductor makes the transition to a normal state. Thermal fluctuations can lead the lattice to melting, then a vortex liquid phase will arise. The melting line is much higher than $H_{c1}(T)$ and below $H_{c2}(T)$. Pinning results in freezing the liquid (near the melting line) into a lattice with a perturbed long-range order, i.e., into the vortex glass phase. If the pinning is weak enough, a narrow tongue of liquid phase enters along the $H_{c1}(T)$ curve from high to low temperatures. Thus the magnetization reversibility in the vicinity of $H_1$ can be ascribed to the presence of the vortex liquid that freezes with increasing field at $H_2$.

### 6.3.2 Temperature Dependence of the Critical Current

As mentioned above, the critical-current value in the crystals was estimated from the dependence of the magnetic-flux penetration depth X on the external field strength $H_c$ (corrected by the demagnetizing factor) [6.74]. Figure 6.4 displays X(H) changes at different temperatures in a Y-Ba-Cu-O crystal where the inhomogeneity is small and X is approximately constant along the sample perimeter. These dependences above the penetration field are seen to be linear. That suggests a linear decrease of induction inside the sample and validates the applicability of the critical-current estimation by $J_c = (c/4\pi)(H_c - H_1)/(1-N)X$. The dependence of $J_c(T)$ thus obtained is presented in Fig.6.5. It is satisfactorily described by $J_c = J_{c0}\exp(-T/T_0)$ with $T_0 = 16$ K, which is in good agreement with data obtained from macroscopic experiments [6.75-79]. The critical-current values observed at $\mathbf{H} \| \boldsymbol{c}$ should be assigned to a current component flowing in the basal plane $J_c^{ab}$. However, due to vortex bending (see below) various current components facilitate field screening, and measured $J_c$ values (at $\mathbf{H} \| \boldsymbol{c}$) give a lower $J_c{}^{ab}$ estimate.[2]

The fast temperature decay observed for the critical current in HTSC single crystals up to now has no strict theoretical interpretation. The strong exponential temperature dependence of $J_c$ is explained for conventional superconductors with a network of Josephson junctions [6.80]. This implies that the granularity of HTSC crystals could be the origin of the dependence Jc(T) (see also [6.18]) and supports the above picture of weak regions revealed in magnetic fields.

---

[2] Improved estimates for $\mathbf{H} \| \gamma$ should account for the demagnetizing effect of superconducting currents.

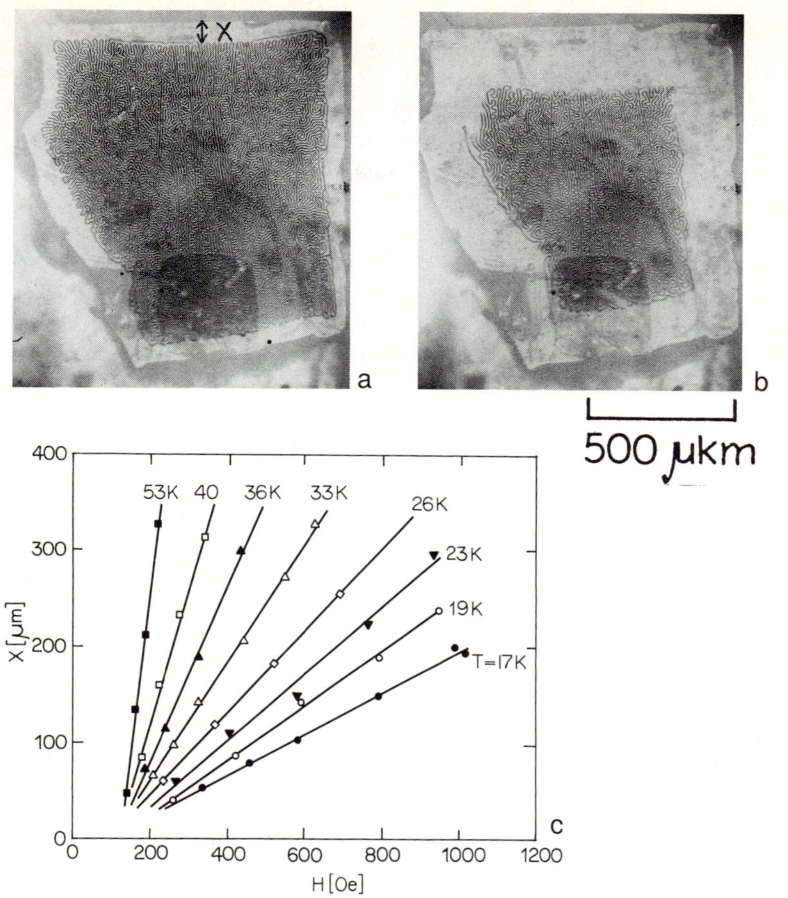


**Fig.6.4.** Displacement of the magnetic flux front with increasing field (H|| *c*). (**a**) and (**b**) show the flux penetration patterns at H = 267 and 578 Oe, respectively (T = 23K); (**c**) distances of the flux front from the sample edge versus the field value at different temperatures

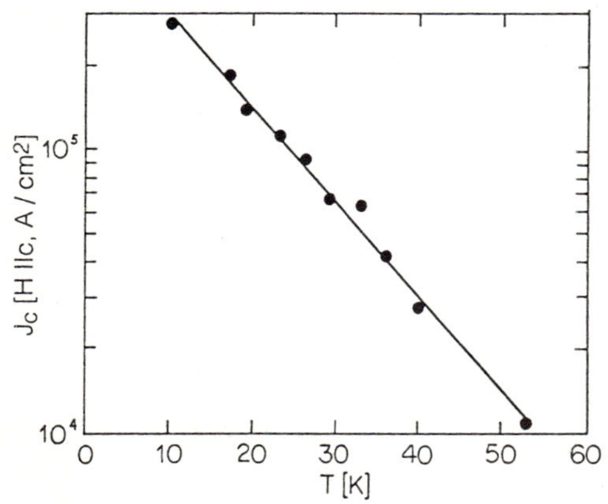


**Fig.6.5.** Temperature changes of the critical current using data of Fig.6.4

An alternative interpretaion associates the $J_c$ decay with the vortex-structure relaxation treated in terms of a Thermally Activated Flux Motion (TAFM), as suggested by, e.g., *Fang* et al. [6.81]. Until recently, attempts at a quantitative treatment of relaxation experiments based on the TAFM description [6.21, 82, 83] using the *Anderson-Kim* model [6.23] yielded non-physical results. Namely, the fundamental parameter of the theory - the pinning potential U - increased with temperature. *Hagen* and *Griessen* [6.5, 84, 85] showed, however, that this problem is resolved if one considers the pinning energy as taking not a single value but obeying some distribution M(U) with a maximum at a given value of U, so that at different temperatures different parts of this distribution contribute to pinning. In the derivation of M(U) especial attention was paid to the substantial role of structure disorder inherent both to ceramics and single crystals. This approach describes a fast increase of magnetic relaxation with temperature and consequently a strong decrease in $J_c(T)$.

An extremely fast critical-current drop, proportional to a high power of 1/T, which in the experiment is hardly distinguished from an exponential decay, is predicted by the collective pinning theory developed by *Feigel'man* et al. [6.86]. In their calculations the effective pinning potential in the random-force model is demonstrated to depend on the superconducting current as $U(J) = U_c(J_c/J)^{\alpha}$. Here $U_c$ is the potential at $J = J_c$, and $\alpha$ is a constant determined by the vortex-lattice dimensionality d and by the dimensionality n of the vector of vortex displacement in the random potential. Then, taking into account that during the time of measurements, t, the critical state is relaxing, the critical current drops and the effective potential becomes $U[J(t)] = T\ell n(t/t_0)$ where $t_0$ is the inverse frequency of attempts of depinning, the current can be written as $J = J_c[U_c/T \times \ell n(t/t_0)]^{1/\alpha}$. For example, in the case of vortex lines (d = 1) able to move in two directions (n = 2) $\alpha = 1/7$ and $J \sim (1/T)^7$ [6.86].

There is also an approach that makes it possible to obtain exactly the exponential $J_c(T)$ dependence. It is based on considerations suggested by *Koshelev* for describing low-temperature changes of the penetration field observed in Tl-cuprate single crystals [6.29]. If experimental observations are conducted during the time t then they reveal a result of the flux creep in this time. Magnetic induction in this period is increased up to some fraction $\delta$ of the external field H. Penetration of vortices through the Bean-Livingston surface barrier was assumed in [6.29] to be the limiting process determining the relaxation time. The barrier height is $U_{BL} = [\Phi_0{}^2 d \times \ell n(H_c/H)]/(4\pi\lambda)^2$, $\Phi_0$ being the flux quantum, $\lambda$ the London penetration depth, $H_c$ the thermodynamic critical field, and d is the vortex length. Then from the creep rate $\delta H/t \sim dB/dt \sim \exp(-U_{bl}/T)$ we find $t = t_0 \exp(U_{BL}/T)$, where $t_0 \simeq (4\pi\delta R\lambda\sigma/c^2\Phi_0)(H_c/H)$ [6.29] with R being the sample dimensions and $\sigma$ the normal-state conductivity. Thus, there is a rather strong dependence t(H) showing that in the time t the action of a definite field (depending on T) will be revealed. This field is obtained by inverting the dependence on t in the form $H = H_c \exp(-T/T_0)$ with $T_0 \simeq \Phi_0{}^2 d/[(4\pi\lambda)^2 \ell n(t/t_0)]$. Then $J_c \sim \Delta M \sim H \sim \exp(-T/T_0)$. Note that $T_0$ is weakly sensitive to the measurement time and sample dimensions entering the logarithm, i.e., $T_0$ can be considered a constant.

In the derivation of the $J_c(T)$ exponential run, the field dependence of pinning potential $U \sim \ln H$ was essential. It is associated with the weak magnetostatic attraction of a vortex to the sample surface. If a similar long-range mechanism (e.g., coupling of vortices to boundaries of weak regions) but not a core interaction determines the flux pinning (at least in low fields) then for such bulk pinning the dependence $J_c \sim \exp(-T/T_0)$ can be derived.

In [6.29] individual 2D vortices were considered, so d was taken to approximately be the distance between CuO layers, and the field dependent value of $t_0$ was estimated as $10^{-10}$ - $10^{-11}$ at H = 1 kOe. Then for t = $10 \div 10^2$ s a value of $T_0$ similar to those observed in HTSCs was obtained. An estimate in [6.29] gave $T_0 \sim 20 \div 25$ K for Tl-cuprate where d ~ 15 Å and $\lambda \sim$ 2000 Å, and $T_0 \sim 10 \div 15$ K for Bi-cuprate with similar d and slightly large $\lambda \sim 3000$ Å.

The characteristic vortex dimension d is important for estimations of $T_0$. To obtain $T_0 \sim 15$ K for Y-Ba-Cu-O using the above model it is necessary to take d of the order of the lattice parameter c, i.e., the existence of 2D vortices in Y-Ba-Cu-O should be assumed. Such a possibility is allowable at ~10 K below $T_c$ [6.6, 87]. However, according to [6.88] the parameter r characterizing the Josephson coupling of a vortex line along the ***c***-direction (the ratio of the characteristic energy for jumps of carriers over the junction between $CuO_2$ layers to $T_c$) for the Y-cuprate is ~2. This indicates a strongly anisotropic but still a 3D vortex behavior to be preferable, although some experimental data do not agree with the 3D description [6.87, 89].

The strong anisotropy which makes the energy of a vortex parallel to the basal plane $E_{ab}$ to be much lower than that of the vortex along the $\gamma$-axis ($E_{ab}/E_c \sim 1/5$ [6.6]) implies that vortex lines can easily bend to the direction of cuprate layers. This bending results in a magnetization component normal to the field direction [6.90]. Then, the vortex penetration and motion under the field **H**||***c*** can be suggested to occur due to formation of loops with segments parallel to the ***ab***-plane - the motion of vortices due to kink formation has already been discussed by e.g., *Iye* et al. [6.91]. If the characteristic loop dimension is on the order of the distance between $CuO_2$ planes, a value of $T_0$ similar to that for Bi- and Tl-cuprates can be obtained; an estimate for Y-Ba-Cu-O gives $T_0 \sim 20$ K at $\lambda = (\lambda_{ab}{}^2 \lambda_c)^{1/3} \sim$ 2000 Å and d ~ 12 Å in accordance with the experiment (the solid line in Fig.6.5 gives $T_0 \sim 16$K).

### 6.3.3 Vortex Bending

The anisotropy of the vortex energy is not the only reason for vortices to bend into the ***ab***-plane. In the case **H**||***c*** this is also induced by the demagnetizing effect resulting in the sample which is enveloped by the force lines. To clarify the role of this factor, the experimental geometry was changed. A sample was mounted with its end face (and the indicator film

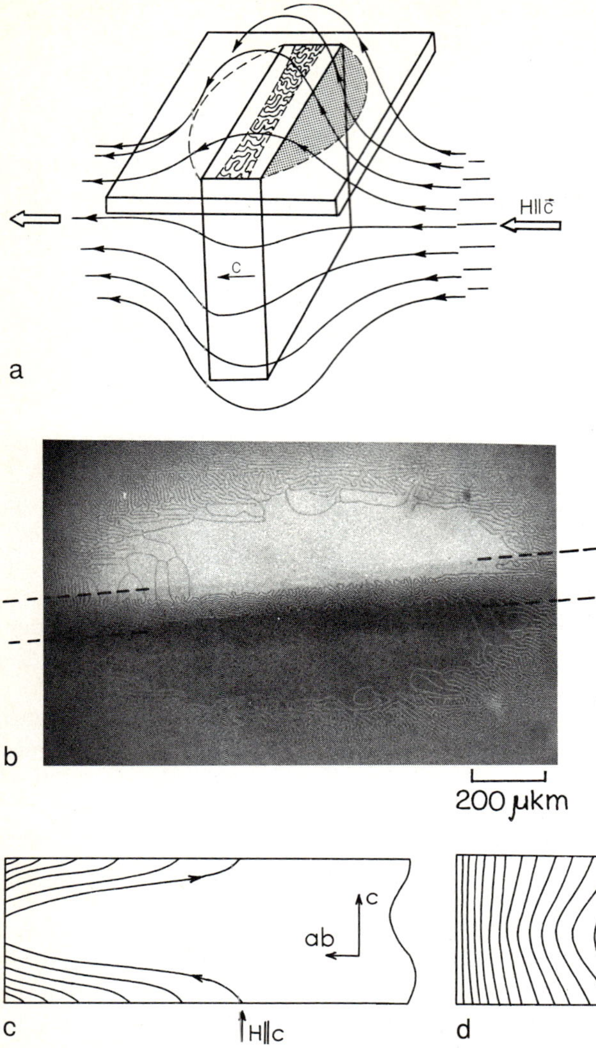


**Fig.6.6.** (**a**) Scheme of the field line distribution when **H**||***c***. (**b**) Magnetic flux penetration pattern at the end face of a Y-Ba-Cu-O single crystal under these conditions (H = 450Oe, T = 30K). (**c**, **d**) changes in the flux line distribution with increasing field

on it) normal to the optical axis of the microscopy; the field was applied normal to the basal plane (**H**||***c***, Fig.6.6a). In this case the uniaxial indicator film is affected only by the induction component normal to the sample's end face, i.e., the vortex projection in the ***ab***-plane. A typical picture is presented in Fig.6.6b, which shows that the penetration of the field **H**||***c*** begins at the sample edges where vortices enter, which are oriented practically perpendicular to **H** and magnetized in opposite directions at different basal surfaces of the plate (Fig.6.6c). In this situation the flux penetration patterns on the basal plane reflect a complicated screening current distribution with the components $J_c^{ab}(H||c)$, $J_c^{c}(H||ab)$ and $J_c^{ab}(H||ab)$ (the superscripts denote the current directions). Thus an evaluation of $J_c$ from the magnetization displacement X observed on a basal plane (Fig.6.5) gives an underestimated $J_c^{ab}(H||c)$ value.

The flux bending, promoted both by the crystal shape and by the anisotropy in the vortex energy, can be the cause of a number of experimentally observed phenomena. First, the vortex bending favors the low-field-magnetization reversibility because the linear tension of the vortices increases the restoring force and the pinning is simultaneously decreased since long vortex sections parallel to the *ab*-plane can easily move along it [6.92]. Second, flux trapping with increasing field is perhaps caused not only by the entry of the vortices into the strong regions but also by the appearance of long vortex sections parallel to *c* when they are moving to the sample center (Fig.6.6d). Obviously during field cooling long flux lines parallel to the $\gamma$-axis are trapped. This should contribute noticeably to the difference of the magnetization patterns observed in zero-field-cooled and field-cooled crystals.

Finally, vortex bending can explain the difference between magnetization pictures in HTSC single crystals and films. In films, as was mentioned, there is practically no reversible stage. Perhaps the reason is that the contribution of flux bending in films is negligible due to their small thickness, and penetrating vortices are mainly oriented along the film normal, which results in their better trapping. Of course, one should keep in mind that the higher critical currents in films, as compared with single crystals [6.57], can correspond to a more effective pinning mechanism by the grain boundaries than by other defects. Note that the vortex bending and the magnetization reversibility associated with it are obviously (together with creep) the reasons for preventing observation of the decoration pattern in HTSC under magnetization, but only after field cooling [6.43-46]

These features were also observed in Tl-cuprate single crystals. But, in addition to the inhomogeneity in the wide basal plane, screening and flux trapping inhomogeneities across the sample thickness were also observed. This inhomogeneity in the form of layers with different properties is probably determined by the presence of plane defects parallel to the basal plane, such as a variation in the number of cuprate layers per cell [6.93] and lattice modulations [6.94] in these crystals. In Bi crystals the flux structure on the face normal to the basal plane was not studied because the quality of the end surfaces did not allow this. However, in these samples, similarly to Tl crystals, a strong flux creep under the field $H||c$ was revealed. The creep manifested itself as a noticeable time variation of the domain structure in the indicator film both after switching off the field in the field-cooling experiment and at a constant field during magnetization of the zero-field-cooled sample. The relaxation was observed even at the lowest temperatures and was much stronger than that for Y-Ba-Cu-O crystals. This observation is explained by a weaker pinning in Bi- and Tl cuprates, also seen by others, e.g., [6.5,95-97].

### 6.3.4 The Critical Current Anisotropy. Magnetic Field in the Basal Plane

Direct evidence of the easy displacement of vortices oriented in the basal plane was given by experiments in which the indicator film was placed on the end face of a crystal and the field was applied normal to this face (**H**|| ***ab***, Fig.6.7a) [6.74]. In this case the field is parallel to the [100] axis but, due to twinning, the ***a***- and ***b***-axes are alternating (in neighbouring twins) along this direction across a crystal, so the index "***ab***" is used.

In the situation described, both a small demagnetizing effect and the vortex-energy anisotropy favor the penetration of straight flux lines parallel to **H**. Therefore, the critical-current anisotropy due to the stronger pinning for vortex motion across the cuprate planes than along them, could be explained.

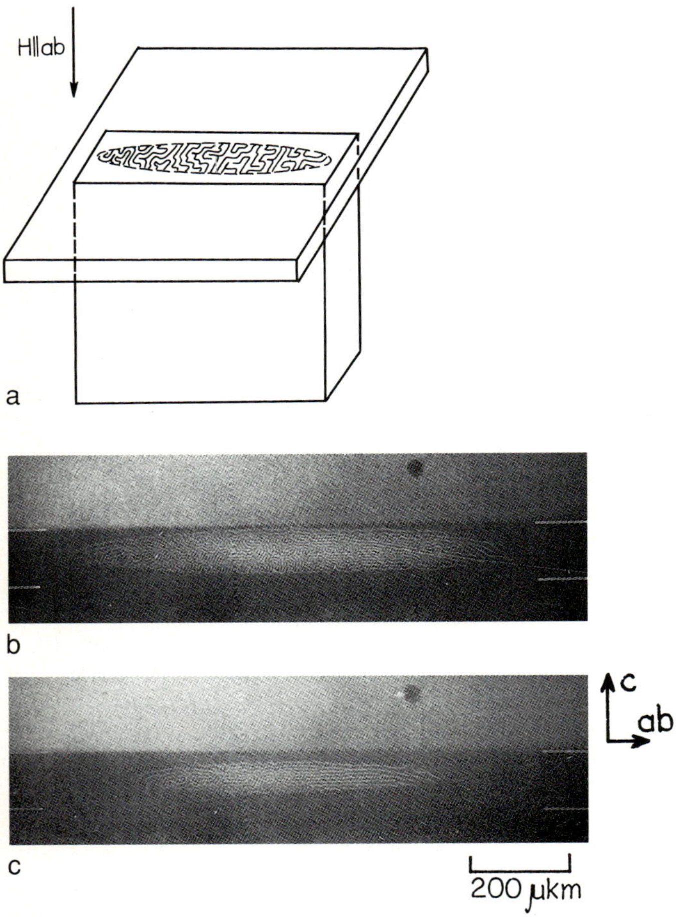


**Fig.6.7.** Magnetic flux penetration in the magnetic field in the basal plane (**H**|| *ab*). (a) scheme of observation; (b) H = 269 Oe, (c) H = 325 Oe, T = 22 K

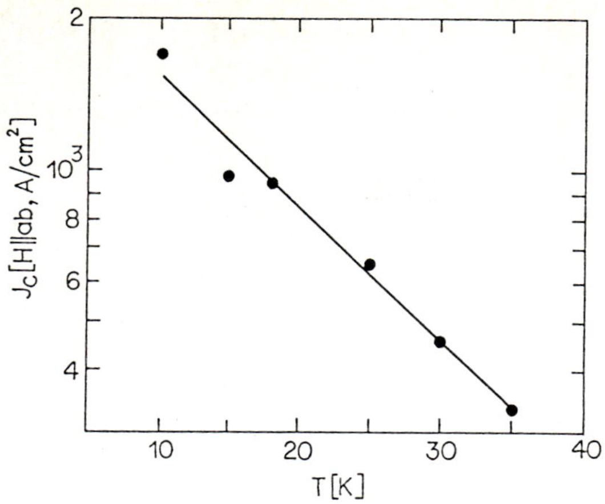


**Fig.6.8.** Temperature dependence of the critical current in the *c*-direction at H||*ab*

As in the **H||*c*** case the penetration of vortices starts at some critical-field value. However, even at low temperatures, the flux-penetration depth differs in value at different edges of the end face. Along the basal plane, vortices enter to a much greater distance than along the plate thickness (i.e., along ***c***). As a result the Meissner region takes the form of an ellipsoid stretched in the ***ab***-plane. The long axis of the ellipsoid is considerably reduced with increasing field, while the short one is decreased only a little (Fig.6.7b, c). By measuring the flux front displacement along the basal plane as a function of the field value, an estimate of the critical current $J_c^c$ (**H||*ab***) was made by using the above discussed formula but with the demagnetizing factor N $\simeq$ 0. Figure 6.8 shows the temperature dependence of $J_c^c$ (**H||*αβ***) for the same crystal for which the critical current along ***ab*** at **H||*c*** is presented in Fig.6.5. It can be seen that $J_c^c$ (**H||*ab***) also decays exponentially against the temperature, the value of $T_0$ = 16 K being close to that for $J_c^{ab}$ (**H||*c***). If the explanation of the exponential $J_c$(T) change in Sect.6.3.2 is valid, then this $T_0$ value could be justified by supposing that vortices parallel to ***ab*** are entering and moving due to the nucleation of kinks with the characteristic dimension d ~ 15 Å.

Note that $J_c{}^c$ (**H||*ab***) turns out to be ~200 times smaller than $J_c^{ab}$ (**H||*c***). As was mentioned, the values of $J_c$ measured by the flux profile at the basal surface in the case **H||*c*** give an underestimate of $J_c{}^{ab}$ (**H||*c***). Thus the pinning for vortices parallel to the ***ab***-plane, moving along the basal plane, is more than two orders of magnitude weaker than that for ***c***-aligned vortices.

Under the experimental conditions it was impossible to measure the flux-front displacement along the crystal thickness to sufficient accuracy and to derive the dependence of $J_c^{ab}$ at **H||*ab*** on T. However, the pinning force estimate from the flux penetration gives a value approximately 7 times larger for the motion of vortices across the cuprate planes than for their displacement along ***ab***. A ratio of the same order ix extracted from macroscopic data for Y-Ba-Cu-O crystals by other researchers [6.10, 53, 63, 92]. But there are also results [6.98-100] showing a ratio, an order of magnitude larger, of the in-plane critical current to the current along the *c*-axis, names ~35 to 59. The in-plane critical current values for both **H**||**c** and **H**||***ab*** have been estimated as nearly equal, e.g., $J_c^{ab}$ (**H**||***c***) = 6.1÷7·10$^5$ and $J_c^{ab}$ (**H**||***ab***) = 4.9·10$^5$ A/cm$^2$ at T = 30 K [6.98]. In our case $J_c^{ab}$ (**H**||***c***) is an order greater than $J_c{}^{ab}$ (**H**||***ab***).

The large values of the critical current anisotropy [6.98-100] obtained from magnetic measurements with **H**||***c*** and **H**||***ab*** seem to agree with the ratio of $J_c^{ab}$ (**H**||***c***) $J_c^{c}$ (**H**||***ab***) derived from our results. It seems that an additional correction accounting for a more realistic flux distribution should be made in the procedure used in [6.98] for extracting the ratio $J_c^{ab}$ (**H**||***c***)/$J_c^{ab}$ (**H**||***ab***) for the two field directions. However, the difference in the anistropy values might also be associated with different defect structures of samples and with the field magnitude in our case being smaller than in the macroscopic experiments (because increasing the field decreases the critical current and reduces the vortex bending at **H**||***c***).

The most natural explanation for the critical current anisotropy observed at **H**||***ab*** is the intrinsic pinning mechanism implying that the lowest energy vortices are those arranged between the $CuO_2$ planes [6.92]. As for the main pinning mechanism for vortices aligned in the ***c***-direction there are different suggestions, from the strong effect of twins [6.10] to the influence of point defects [6.46]. This problem will be discussed in more detail below.

### 6.3.5 Effects of Twins on the Magnetic Properties of 1-2-3 Cuprates

Twins are an integral component of the structure of orthorhombic HTSC compounds. Untwinned crystals are obtained only by using special tricks: by heating them under a uniaxial load, as done in, e.g., [6.59, 100-102], or by quenching samples from the tetragonal phase and then oxygenating them [6.103]. Two principal questions concerning the effects of twin boundaries are raised as mentioned above. First, is the superconductivity increased or decreased at the twin boundaries? Second, what is the contribution of twins to the magnetic flux pinning?

To clarify these problems the local superconducting parameters (like $T_c$, penetration field, critical current) were studied in as-grown Y-Ba-Cu-O single crystals possessing strongly inhomogeneous twin structure [6.57]. In addition, a comparison of these characteristics before and after untwinning was made [6.59].

The twin structure of samples was analyzed in polarized light employing the bireflection effect [6.60-62]. Twin walls aligned in the ⟨110⟩ directions were easily revealed when the domain width D was larger than the half wavelength $\lambda/2$. The contrast of domains was reversed by rotating the sample with respect to the microscope polarizers; this enabled determination of the directions of the ***a*** and ***b***-axes in the domains [6.61]. In the case D < $\lambda/2$, separate domains were practically invisible (the minimum domain width which could be resolved using an immersion objective was ~0.1 $\mu$m) and only variations in the twin density were observed.

An investigation of twin-motion processes under a concentrated load [6.104] showed that the mobility of twin boundaries, determined by the oxygen-vacancy diffusion and described in terms of the displacement of partial twinning dislocations, was rapidly increased with rising temperature. However, at sufficiently high temperature, a large escape of oxygen from the crystals occurred that was accompanied by changes in the phase composition [6.105-106]. Thus, for untwinning samples under the action of a uniaxial loading, a temperature $T < 300$ K was chosen, and the experiment was conducted in an oxygen atmosphere [6.59].

In most of the as-grown crystals where regions with considerably different twin densities existed, a correlation between the twin structure and local superconducting properties was revealed. First of all, in densly twinned regions the penetration field was noticeably lower than that in regions with wide twin domains. Figure 6.9 exhibits a typical magnetization picture on an example of a Y-Ba-Cu-O crystal with wide domains in one corner and with high, optically unresolved, twin density in the rest of the sample volume (Fig.6.9a). The presence of twins there is checked by identation resulting in the formation of the detwinning rosette [6.104].

A better screening of the external field is observed in the region with lower twin density, whereas in the surrounding area the flux enters much more easily (Fig.6.9b). After field cooling the flux trapping is also stronger in the corner (Fig.6.9c). Local transition curves measured in the appropriate regions of the sample using the modulation method show that in the strong screening area there is a higher $T_c$ (Fig.6.10). Note that the macroscopic susceptibility curve is consistent with the $T_c$ inhomogeneity in the crystal, as indicated by the $\chi(T)$ curve in the same figure.

This inhomogeneity in the superconducting characteristics could be considered as evidence of weak-link formation at the twin boundaries, which was discussed in [6.15], and as an argument against increasing superconductivity there, assumed in [6.16, 17, 107]. However, the untwinning experiment showed that the twin boundary effect is not the determining factor here [6.59].

Figure 6.11a displays local transition curves at points with different as-grown twin density in a crystal where the effect of detwinning was studied. The transition temperature changes as one moves from the sample edge with wide domains, where a strong screening was found in accordance with above-described observation (Fig.6.11e), into the densely twinned area. Three distinct phases can be identified with $T_c \simeq 90$, 70 and 55 K, respectively. The transitions are fairly wide indicating that these phases are inhomogeneous.

After detwinning the crystal, most of it became a single domain and a partial decomposition of the 70 K phase took place. (At point 5 shown in Fig.6.11e this phase was then not revealed at all, and at point 2 the transition widened indicating the formation of the $T_c = 55$ K phase, Fig. 6.11b).

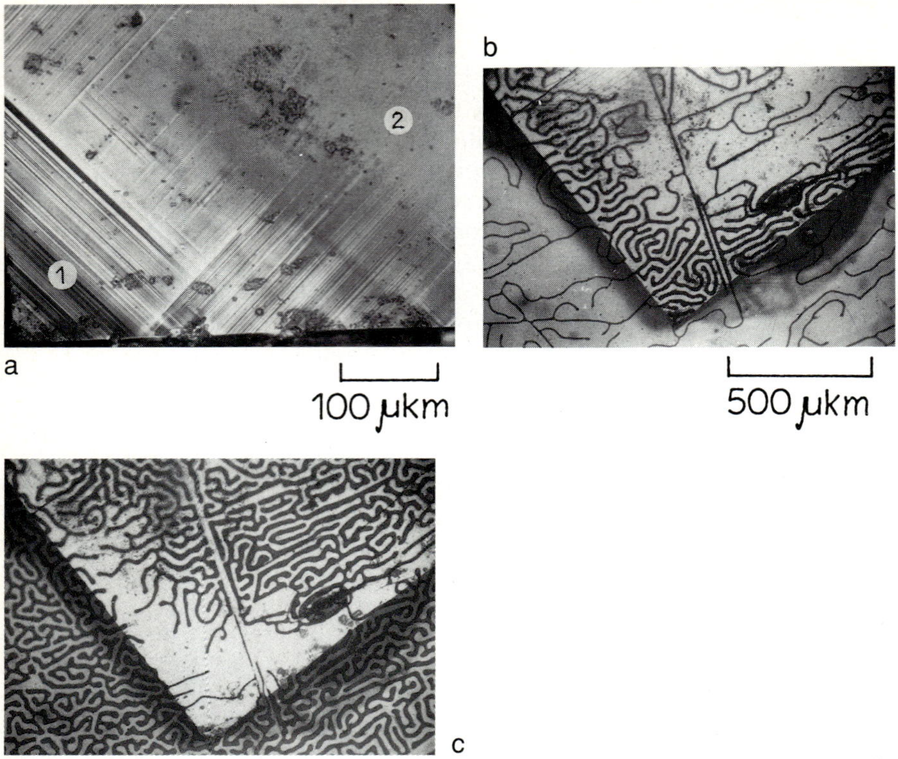


**Fig.6.9.** Features of penetration and trapped flux patterns in an as-grown Y-Ba-Cu-O crystal with inhomogeneous twin structure. (**a**) In the corner of the sample twin domains are revealed in polarized light, and except for in this corner twins are not resolved due to their high density; (**b**) better screening (H = 60Oe, T = 14K), and (**c**) stronger flux trapping (after cooling in a 76Oe field) (H = 0, T = 14K) are observed in the region with wider domains. The white spots *1* and *2* indicate the regions wher the local susceptibility curves shown in Fig.6.10 were measured

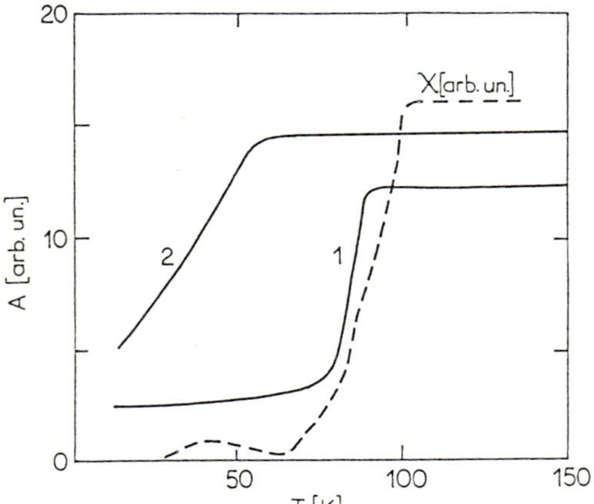


**Fig.6.10.** Local susceptibility curves in regions with wide (*1*) and fine (*2*) twins (as indicated in Fig.6.9a). The curves are derived by measuring the amplitudes A of domain wall vibrations in the indicator film above a chosen point of the crystal in a small AC field in the *c*-direction. The *dashed line* is the macroscopic susceptibility curve of the sample

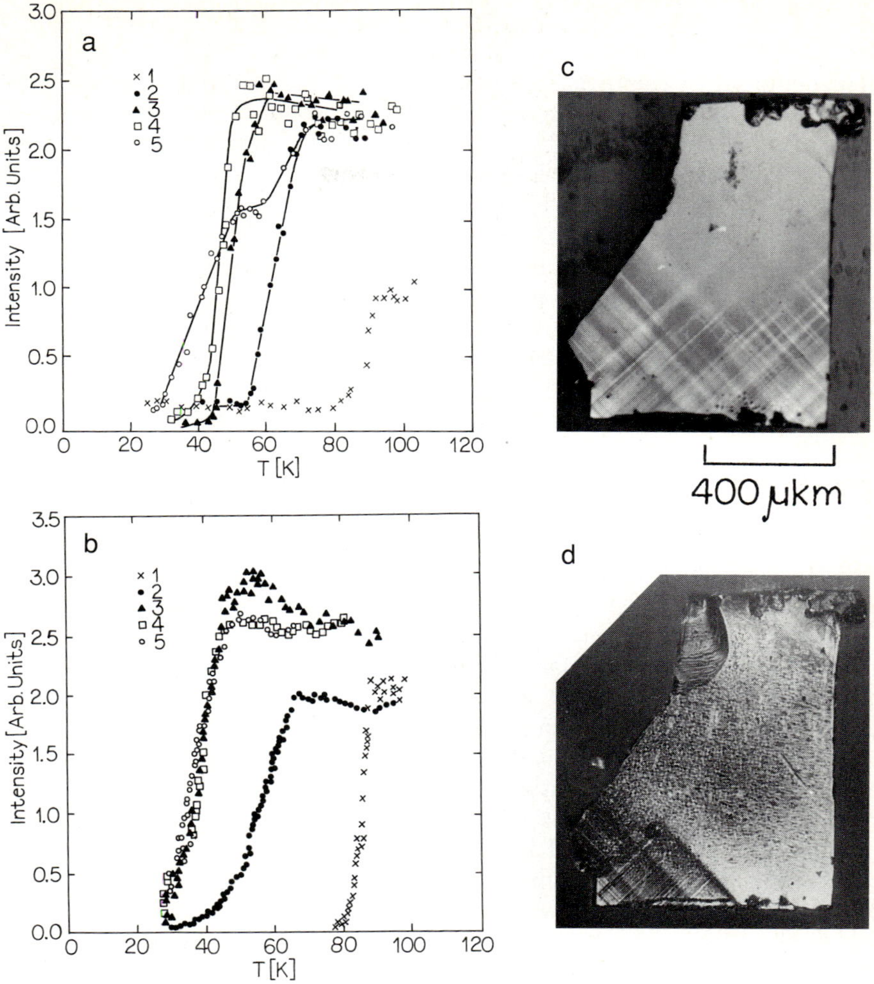


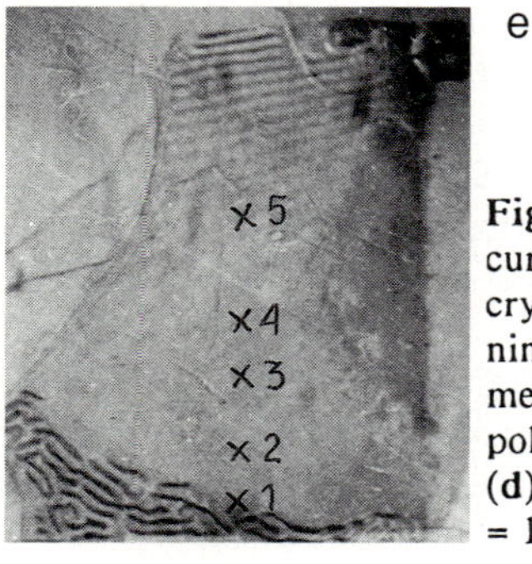


**Fig.6.11.** Comparison of the local susceptibility curves in an inhomogeneous Y-Ba-Cu-O single crystal measured before (**a**) and after (**b**) detwinning the sample. (Points where the curves were measured are indicated in part **e**.) Pictures in the polarized light of the crystal before (**c**) and after (**d**) detwinning; (**e**) field penetration pattern at H = 160 Oe T = 26 K in the detwinned crystal

These changes in the phase composition seem to be associated with a short-time heating of the sample and not with the disappearance of the twin boundaries. They were also observed in a control sample placed near the detwinned one on the heating table but not subjected to the uniaxial stress. Taking into account the known relation between the superconducting transition temperature and the oxygen content in Y-Ba-Cu-O [6.31] the phase change could be due to the oxygen redistribution during heating of the crystal. Note that at point 1 where the highest $T_c$ is observed and where the oxygen content Z is accordingly expected to be maximum, the effect of heating during detwinning on the transition curve is minimum. But in regions with lower Z (and lower $T_c$, respectively) changes are more noticeable. This just corresponds to what is known about the oxygen redistribution and ordering in chains [6.108].

It is significant that despite the detwinning, the general picture of the transition-temperature distribution was not changed. In addition, the field penetration pattern was changed only insignificantly: the strong screening region remained at the same place and the values of the penetration field in different areas were for the most part the same. This means that the role of twins in the determination of superconducting properties is perhaps less essential than the contribution of composition inhomogeneities.

Variations of $T_c$, of the penetration field, and of pinning forces in 1-2-3 crystals can be associated first of all with the presence of a weak-link network resulting from the decomposition of the initial phase into new phases with different oxygen content. As the superconducting transition in this case should depend strongly on the external field value, the local susceptibility curves obtained using the modulation method should shift appreciably to lower temperatures with increasing field amplitude $H_0$. Precisely such an effect was observed in regions with low $T_c$ and easy flux penetration. In the 90 K phase regions the transition onset practically did not shift and only its width was somewhat increased with $H_0$.

Such a difference is obviously connected with the formation of a continuous weak-link network in regions with lower oxygen content, whose effect becomes dominating. This is possible when extended weak links are separated by a distance smaller than the magnetic penetration depth. In regions of hard penetration, magnetic properties are, evidently, mainly determined by the bulk 90 K phase.

After detwinning the effect of the field amplitude on the transition was not changed. This is another indication that other defects (such as layers of oxygen-depleted phases) are more effective weak links in HTSC crystals than twin boundaries.

Thus the correlation of the twin boundary density and superconducting properties observed in inhomogeneous as-grown HTSC crystals should perhaps be related to the influence of the oxygen content on the twin size. Since the orthorhombicity of 1-2-3 compounds is decreased with decreasing concentration of the O4 atoms, the energy of formation of twins should also decrease and their density should accordingly increase. A similar effect was observed experimentally by *Van Tendeloo* et al. [6.109]. In addition, accounting for the dependence of the twin width D on the crystal thickness L (it can be shown that D ~ L) a refinement of twins in regions with lower $T_c$ may be associated with oxygen entering only a thin surface crust (in regions far enough away from the sample edges) where the orthorhombic superconducting phase is just formed. Both reasons indicate that twins must be wider in regions with larger oxygen content where superconducting properties are higher, independently of the twin structure. Note that in the corners of thick as-grown crystals where the oxygen entry is facilitated, wide twins, and higher $T_c$ and $J_c$ values are observed most frequently.

A confirmation of the insignificant twin-boundary effect on $T_c$ is also given by studies of magnetization processes and measurements of local transition curves in samples with lower composition inhomogeneity. These crystals have comparatively narrow macroscopic transition curves and, as a rule, twins are optically well resolved in them. Often they have wide single-domain regions. However, no correlation between the twin density and the local-transition curve was observed in such samples. For example, in a region with domain width $D < 0.2\ \mu m$ in one of the samples, the local transition curve was the same as in a twin-free area. Widening of the transition with increasing modulating field amplitude did not vary noticably across the crystal.

Moreover, at sufficiently low temperatures the flux-penetration picture was not connected with features of the twin-domain structure. As mentioned above, at $T < 20$ K the flux-penetration front was following the crystal contour, which corresponds to a homogeneous pinning on uniformly distributed defects. With increasing temperature, however, some pinning inhomogeneites that were not bound to twins, and perhaps were caused by some composition changes, began to show.

Beginning from $T > 30$ K, the preferential magnetic-flux penetration along the twin-boundary direction could be observed (Fig.6.12). This observation demonstrates clearly that only at high enough temperatures can twin boundaries become the most effective pinning centers preventing vortices from moving across their planes.

Theoretically the question about the significance of twin boundaries for pinning in orthorhombic HTSC crystals can be solved unambiguously [6.10, 17]. A pinning of vortices at a twin boundary is predicted, independent of increasing or decreasing superconductivity in the vicinity of a twinning plane, having repulsive or attractive interaction potential, respectively [6.110]. In many experimental works the effect of twins on the pinning was really revealed in 1-2-3 crystals. First of all it was determined from measurements of macroscopic magnetization curves and of the Meissner fraction in different directions of the field. For example, *Welp* et al. [6.100] revealed a decrease of the M(H) hysteresis after detwinning a Y-Ba-Cu-O single crystal. The effect was strongest at 10 K when the field was oriented in the basal plane along the direction of the initial twin walls. In the case $\mathbf{H} \| \boldsymbol{c}$ a weaker effect was noted at $T > 65$ K. *Kartsovnik* et al. [6.40] observed some decrease of the Meissner fraction in the range from 20 K to $T_c$ in a Y-Ba-Cu-O sample with a preferential twin direction when the field was applied

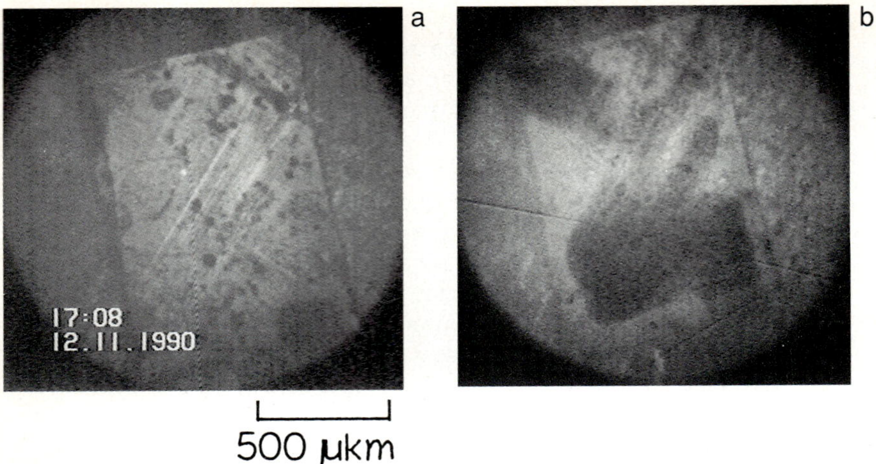


Fig.6.12. Effect of twins on the flux penetration. (a) Twin domains in a Y-Ba-Cu-O single crystal; (b) penetration pattern at H = 380 Oe, T = 39 K; *darker areas* are Meissner regions; flux entering (*brighter areas*) along the twin boundaries is seen

in the ***ab***-plane along the twin boundaries boundaries. *Keller* et al. [6.111] measured the M(H) hysteresis at T > 75 K in oriented Y-Ba-Cu-O ceramics and established that the volume pinning force in the field **H**||***c*** (parallel to twin planes) is an order of magnitude higher than that for **H**||***ab*** (not along twin boundaries). A very small but measurable effect of twins on the pinning was observed by *Swarzendruber* et al. [6.39] who measured magnetization curves at 10 K in a Y-Ba-Cu-O crystal with a single twin direction. The effect was became obvious in the field parallel to the basal plane as some difference in M(H) curves for **H** parallel and **H** perpendicular to the twin boundaries.

Although indirect, highly convincing evidence of the pinning by twins was given by transport measurements in Y-Ba-Cu-O crystals with one and two twinning directions by *Kwok* et al. [6.112]. Narrow maxima in resistance curves as functions of the field direction (at $T \sim T_c$) were revealed when the field parallel to the ***ab***-plane was oriented along twin boundaries. Similar conductivity measurements by *Iye* et al. [6.113] in Y-Ba-Cu-O films near $T_c$ support the result of [6.111]. *Roas* et al. [6.114], studying the transport of critical current for different field directions, in addition to strong pinning by $CuO_2$ planes, also observed some pinning increase when the field was along the ***c***-axis, which was associated with the effect of twin boundaries.

Finally, proof for pinning by twins was obtained from mechanical torque measurements. Torque cusps due to pinning were found in Y- and Gd-cuprate crystals at 77 K by *Hergt* et al. [6.115] when the field direction coincided with the plane of twin boundaries. *Janossy* et al. [6.116], conducting similar studies in Y-Ba-Cu-O crystals in a wide temperature range, observed pinning in the field **H**||***c***, ascribed to twins, which had a maximum strength at 25 K and decayed rapidly with decreasing temperature to 10 K and also with increasing T.

Observation of the powder accumulation at twin boundaries in decoration experiments [6.45, 46] could also be cited as proofs that the boundaries are effective centers of pinning. However, there is a question of whether or not they serve as weak channels for flux penetration and disappear when cooling the crystal in the field.

Along with the above results, there are numerous experimental data indicating the insignificant role of twins in the pinning. For example, *Shi* et al. [6.117] observed pinning weakening after refinement of the twin structure (although this was achieved by annealing samples, which also resulted in decreasing $T_c$). They noted that channels of easy flux entry could be formed at the twin boundaries by accumulation of oxygen vacancies in their vicinity. *Blunt* et al. [6.118] decreased the twin domain width by doping Y-Ba-Cu-O crystals with Co and did not see any change in the critical-current value extracted from M(H) curves at 4.2 K. Then they concluded that pinning by twins is not significant. *Gyorgy* et al. [6.98] calculated different current components from M(H) curves taking into account the anisotropy of Y-Ba-Cu-O crystals and found that critical currents flowing in the basal plane both at $\mathbf{H} \| \boldsymbol{c}$ and $\mathbf{H} \| \boldsymbol{ab}$ have similar values. Thus, taking into account the different directions of vortex motion for these two orientations of **H**, *Gyorgy* et al. [6.98] proposed that the effect of twins on pinning is small (at least at the experimental temperature of ~30K).

After detwinning a Y-Ba-Cu-O crystal *Liu* et al. [6.119] observed at T = 6 only a small increase of the Meissner fraction. The invariability of the critical current at $\mathbf{H} \| \boldsymbol{c}$ and a weak $J_c$ decrease at $\mathbf{H} \| \boldsymbol{ab}$ were also revealed. That led *Liu* et al. to suggest that twin boundaries are not the main pinning centers.

Observing the field penetration in a Y-Ba-Cu-O crystal at T = 10 K using the magnetooptic covering technique, *Koblischka* et al. [6.120] also noted a weak influence of twins on the flux distribution.

Summarizing the data of these different groups we can conclude that a maximum pinning on the twin boundaries takes place when vortices are lying along them in the basal plane. At higher temperatures a pinning on twins is also revealed when $\mathbf{H} \| \boldsymbol{c}$. With decreasing temperature below ~30 K other pinning centers become more effective. Note that this picture also explains our results described above.

The pinning mechanism in HTSCs, which dominates at low temperatures, is still poorly understood. Suggestions have been made that it can be an intrinsic pinning on the crystal lattice itself [6.46, 120], related to the very small size of the vortex cores, explaining the sensitivity of the latter to their positions with respect to the lattice. Such a mechanism was considered by *Shimmele* [6.121] within the collective pinning model. In addition, due to the small coherence length, point defects (such as oxygen vacancies in $CuO_2$ planes and impurities) and CuO chains with a deficiency of oxygen [6.10, 46, 122] should be effective pins.

Introduction of dislocations into HTSC crystals showed that they can also be strong pinning centers [6.115]. However, it seems doubtful that there should be many dislocations in as-grown crystals.

*Shi* et al. [6.123] noted that in the temperature range $30 < T < 70$ K the pinning could be determined by the inhomogeneity in the oxygen content, resulting in variations in the superconducting parameters and thus in the appearance of a potential relief for the vortex motion. Variations in the oxygen content in the vicinity of twin boundaries is also considered a reason for pinning by twins [6.46, 100].

The inhomogeneous oxygen distribution that should manifest itself primarily in the variation of the concentration of the oxygen vacancies in CuO chains (and, generally speaking, can be accompanied by a phase stratification of HTSC crystals) might be the main factor influencing the flux-line motion. At low temperatures, vacancies in the chains (and maybe their clusters) are obviously dominating anchors for Abrikosov vortices. With increasing temperature, the regions with lower oxygen concentration can begin to serve as the easy flux-penetration channels, and twins become more effective pinning centers. For a more detailed elucidation of the mechanisms of vortex trapping in HTSC crystals, a time-cosuming experimental work is required.

## 6.4 Conclusions

The method for visualization of the magnetic flux in superconductors using iron-garnet indicator films reveals inhomogeneities of superconducting properties in HTSC single crystals and films: regions of higher and lower local transition temperature $T_c$, the penetration (or first critical) field, and the critical current $J_c$ values are derived.

The magnetic field penetration process in crystals shows two stages: a reversible and then an irreversible one. The first stage can be connected with vortices entering along "weak" channels, with insignificant pinning. The second one seems to correspond to penetration of flux lines into "strong" regions of noticeable pinning. In the field $\mathbf{H} \| \boldsymbol{c}$ considerable vortex bending was revealed associated with strong anistropy of the vortex energy and the demagnetizing effect. This bending can also contribute to the low-field magnetization reversibility.

In agreement with the data of many others, our results show a strong exponential temperature decay of the critical current in HTSCs. It is determined by the thermally activated flux motion. If the vortex motion occurs after overcoming a long-range (but not core) pinning potential (e.g., for a vortex moving in a weak channel, the repulsive potential from a strong region) by flux-line loops with a characteristic size on the order of the distance between $CuO_2$ layers, then both the exponential $J_c(T)$ change and the value of $T_0$ in the exponent coinciding with the experimental values can be derived.

The critical-current anisotropy is determined by observing the field penetration on different faces of Y-Ba-Cu-O crystals in various field directions. The critical current along the *c*-axis is approximately 7 times lower than $J_c$ in the basal plane at $\mathbf{H} \| \boldsymbol{ab}$. The latter is more than an order of magnitude less than the critical current in the plane at $\mathbf{H} \| \boldsymbol{c}$. The above ratios indicate that minimum pinning takes place when vortices parallel to the cuprate planes move along them.

In as-grown Y-Ba-Cu-O crystals a correlation is observed between increasing twin density and decreasing superconducting properties. However, it is established that this correlation is associated with a variation of the twin width, which is dependent on the oxygen content, a decrease of which determines decreasing $T_c$, $H_{c1}$ and $J_c$.

At low temperatures the role of twins in the pinning is insignificant. The intrinsic lattice pinning and pinning on randomly distributed point defects (perhaps on oxygen vacancies in CuO chains) are probably dominating. However, at $T > 30$ K, twins become effective pins for motion of vortices across their boundaries.

In conclusion, we note that the developed method for observing magnetic flux in superconductors is a simple and effective tool for the quality control of high-$T_c$ samples.

## References


6.1 A.P. Malozemoff: Macroscopic magnetic properties of high temperature superconductors, in *Physical Properties of High Temperature Superconductors I*, ed. by D.M. Ginsberg (World Scientific, Singapore 1989) pp.71-150

6.2 E.Z. Meilikhov: Sverkhprovodimost (KIAE) 2, 5-29 (1989). [Engl. transl.: Superconductivity 2, 7 (1989)]

6.3 U. Welp, W.K. Kwok, G.W. Grabtree, K.G. Vandervoort, J.Z. Liu: Physica C **162-164**, 735-736 (1989)

6.4 L. Krusin-Elbaum, A.P. Malozemoff, Y. Yeshurun, D.C. Cronmeyer, F. Holzberg: Phys. Rev. B **39**, 2936-2939 (1989)

6.5 C.W. Hagen, R. Greissen: Thermally activated magnetic relaxation in high $T_c$ superconductors, in *Studies of High Temperature Superconductors* ed. by A.V. Narlikar, Vol.3 (Nova Science, Commack, NY 1989) pp.159-195

6.6 P.M. Kes: In *Transport Properties of Superconductors, Progress in High Temperature Superconductivity*, ed. by R. Nikolsky (World Scientific, Singapore 1990) pp.119-125

6.7 W.E. Lowrence, S. Doniach: In *Low Temperature Physics LT 12*, ed. by E. Kanda (Keigaku, Tokyo 1971) p.361

6.8 L.N. Bulaevskii: Zh. Eksp. Teor. Fiz. **64**, 2241-2247 (1973) [Engl. transl.: Sov. Phys.-JETP **37**, 1133 (1973)]

6.9 A.P. Malozemoff, T.K. Worthington, E. Zeldov, N.C. Yeh, M.W. McElfresh, F. Holzberg: Flux creep and the crossover to flux flow in the resistivity of high-$T_c$ superconductors, in *Strong Correlation and Superconductivity*, ed. by H. Fukuyama, S. Maekawa, A.P. Malozemoff, Springer Ser. Solid-State Sci., Vol.89 (Springer, Berlin, Heidelberg 1989) pp.349-360

6.10 P.H. Kes: Physica C **153-155**, 1121-1126 (1988)

6.11 J. Clem: Physica C **153-155**, 50-55 (1988)

6.12 K.A. Müller, M. Takashige, J.G. Bednorz: Phys. Rev. Lett. **58**, 1143-1146 (1987)
K.A. Müller, K.W. Blazey, J.G. Bednorz, M. Takashige: Phys. Rev. B **148**, 149-154 (1987)
6.13 M. Daemling, J. Seuntjens, D.C. Larbalestier: Nature **346**, 332-335 (1990)
6.14 H. Küpfer, I. Apfelstedt, R. Flikiger, C. Keller, R. Meier-Hirmer, B. Runtsch, A. Turowski, U. Wiech, T. Wolf: Cryogenics **29**, 268-280 (1989)
C. Keller, H. Küpfer, R. Meier-Hirmer, U. Wiech, V. Selvamanickam, K. Salama: Cryogenics **30**, 401-409 (1990)
6.15 G. Deutscher, K.A. Müller: Phys. Rev. Lett. **59**, 1745-1747 (1987)
6.16 I.N. Khlyustikov, A.I. Buzdin: Adv. Phys. **36**, 271-330 (1987)
6.17 A.A. Abrikosov, A.I. Buzdin, M.L. Kulic, D.A. Kuptsov: Sci. Techol. **1**, 260-268 (1989)
6.18 M.M. Fang, V.G. Kogan, D.K. Finnemore, J.R. Clem, L.S. Chumbley, D.E. Farell: Phys. Rev. B **37**, 2334-2337 (1988)
6.19 S.E. Inderhees, M.B. Salamon, N. Goldenfeld, J.P. Rice, B.G. Pazol, D.M. Ginsberg: Phys. Rev. Lett. **60**, 1178-1180 (1988)
6.20 R.A. Buttera: Phys. Rev. B **37**, 5909-5911 (1988)
6.21 Y. Yeshurun, A.P. Malozemoff: Phys. Rev. Lett. **60**, 2202-2205 (1988)
6.22 P.W. Anderson: Phys. Rev. Lett. **9**, 309-311 (1962)
6.23 P.W. Anderson, Y.B. Kim: Rev. Mod. Phys. **36**, 39-43 (1964)
6.24 D.R. Nelson: Phys. Rev. **60**, 1973-1976 (1988)
6.25 M.P.A. Fisher: Phys. Rev. Lett. **62**, 1415-1418 (1989)
6.26 D.S. Fisher, M.P.A. Fisher, D.A. Huse: Thermal fluctuations, quenched disorder, phase transitions and transport in type-II superconductors, preprint (1990)
6.27 A.P. Malozemoff, T.K. Worthington, R.M. Yandorofski, Y. Yeshurun: In *Towards the Theoretical Understanding of High Temperature Superconductors*, ed. by S. Lundgvist, E. Tosatti, Yu. Lu (World Scientific, Singapore, 1988)
6.28 A.P. Malozemoff, T.K. Worthington, Y. Yeshurun, F. Holzberg: Phys. Rev. B **38**, 723-7207 (1988)
6.29 V.N. Kopylov, A.E. Koshelev, I.F. Schegolev, T.G. Togonidze: Physica C **170**, 291-297 (1990)
6.30 K.N. Tu, C.C. Tsuei, S.I. Park, A. Levi: Phys. Rev. B **38**, 772-775 (1988)
N.C. Yeh, K.N. Tu, S.I. Park, C.C. Tsuei: Phys. Rev. B **38**, 7087-7090 (1988)
S.J. Rothmann, J.L. Routbort, J.E. Barker: Phys. Rev. B **40**, 8852-8860 (1989)
6.31 R.J. Cava, B. Battlogy, C.H. Chen, E.A. Reitman, S.M. Zahurak, D.J. Werder: Phys. Rev. B **36**, 5719-5722 (1987)
D.J. Werder, C.H. Chen, R.J. Cava, B. Battlogy: Phys. Rev. B **37**, 2317-2319 (1988)
6.32 S.I. Park, C.C. Tsuei, K.N. Tu: Phys. Rev. B **37**, 2305-2308 (1988)
6.33 E.H. Brandt, U. Essmann: Phys. Stat. Sol. (b) **144**, 13-38 (1987)
6.34 P.L. Gammel, L.F. Schneemeyer, J.V. Waszcak, D.J. Bishop: Phys. Rev. Lett. **61**, 1666-1669 (1988)
6.35 D.E. Farell, C.M. Williams, S.A. Wolf, N.P. Bansal, V.G. Kogan: Phys. Rev. Lett. **61**, 2805-2808 (1988)
6.36 B. Pümpin, H. Keller, W. Kündig, W. Odermatf, I.M. Savic, J.W. Schneider, J.G. Bednorz, Y. Maeno, K.A. Müller, C. Rossel, E. Kaldis, S. Rusiecki, W. Assmus, J. Kovalevski: Physica C **162-164**, 151-152 (1989)
6.37 S.G. Barsov, A.A. Vasiliev, A.L. Getalov, V.P. Koptev, S.M. Kozlov, L.A. Kusmin, S.M. Mikirtychyants, R.K. Nikolaev, N.S. Sidorov, V.K. Fedotov, E.P. Fedotova-Koval, G.V. Sherbakov: Sverkhprovodimost (KIAE) 2, 42-50 (1989) [Engl. transl.: Superconductivity (1989)]

6.38 L. Fruchter, G. Fillon, I.A. Campbell: Proc. Eur. Conf. of High-$T_c$ Thin Films and Single Crystals (Ustron, Poland, 1989)

6.39 L.J. Swartzendruber, A. Roitburd, D.L. Kaiser, F.W. Gayle, L.H. Bennet: Phys. Rev. Lett. **64**, 483-486 (1990)

6.40 M.V. Kartsovnik, G.Yu. Logvenov, K.Ya. Soifer: Cryogenics **30**, 647-649 (1990)

6.41 B.M. Balashova, Yu.V. Sharvin: Zh. Eksp. Teor. Fiz. **31**, 40-44 (1956) [Engl. transl.: Sov. Phys.-JETP **4**, 54 (1957)]

6.42 V. Essmann, H. Träuble: Phys. Lett. A **24**, 526-527 (1967)
H. Träuble, U. Essmann: Phys. Stat. Sol. **25**, 373-393 (1968)

6.43 P.L. Gammel, D.Z. Bishop, G.J. Dolan, J.R. Kwo, C.A. Murray, L.F. Schneemeyer, J.V. Waszczak: Phys. Rev. Lett. **59**, 2592-2595 (1987)

6.44 L.Ya. Vinnikov, L.A. Gurevich, G.A. Emelchenko, Yu.A. Ossipyan: Pisma Zh. Eksp. Teor. Fiz. **47**, 109-111 (1988) [Engl. transl.: Sov. Phys.-JETP Lett. (1989)]

6.45 L.Ya. Vinnikov, L.A. Gurevich, G.A. Emelchenko, Yu.A. Ossipyan: Sol. State Commun. **67**, 421-423 (1988)

6.46 G.J. Dolan, G.V. Chandrashekhar, T.R. Dinger, C. Feild, F. Holzberg: Phys. Rev. Let. **62**, 827-830 (1989)

6.47 K. Osamura, Y. Tomii, N. Matsukura, Sh. Ochiai: Jpn. J. Appl. Phys. **27**, L860-L863 (1988)

6.48 H.F. Hess, R.B. Robinson, R.C. Dynes, J.M. Valles, J.V. Waszczar: Phys. Rev. Lett. **62**, 214-216 (1989)

6.49 H.U. Habermeier, R. Aoki, H. Kronmüller: Phys. Stat. Sol. (a) **50**, 187-194 (1978)

6.50 N. Moser, M.R. Koblischka, B. Gegenheimer, H. Kronmüller, H. Theuss: Physica C **159**, 117-123 (1989)

6.51 E. Batalla, R. Goudreaulf, E.G. Zwartz, B.A. Judd, L.S. Wright: in *Physics and Material Science of High Temperature Superconductors*, ed. by R. Kossowsky, S. Methfessel, D. Wohlleben (Kluwer Academic, Dordrecht 1990) pp.417-427

6.52 A.A. Polyanskii, V.K. Vlasko-Vlasov, M.V. Indenbom, V.I. Nikitenko: Pisma Zh. Tekh. Fiz. **15**, 1-4 (1989) [Engl. transl.: Sov. Phys.-Tech. Phys. Lett. 15: 872-873 (1989)]

6.53 S. Gotoh, N. Koshizuka, M. Murakami, S. Tanaka: Jpn. J. Appl. Phys. **29**, 1083-1085 (1990)

6.54 R. Szymczak, K. Piotrowski, H. Szymczak, A.M. Balbashov: J. Magn. Magn. Mater. **92**, L19-L24 (1990)

6.55 A.A. Polyanskii, V.K. Vlasko-Vlasov, M.V. Indenbom, V.I. Nikitenko: In *Physics and Material Science of High Temperature Superconductors*, ed. by R. Kossowsky, S. Methfessel, D. Wohlleben (Kluwer Academic, Dordrecht 1990) pp.131-139

6.56 M.V. Indenbom, N.N. Koleshnikov, M.P. Kulakov, I.G. Naumenko, V.I. Nikitenko, A.A. Polyanskii, N.F. Vershinin, V.K. Vlasko-Vlasov: Physica C **166**, 486-496 (1990)

6.57 V.K. Vlasko-Vlasov, M.V. Indenbom, V.I. Nikitenko, Yu.A. Ossipyan, A.A. Polyanskii, R.L. Prozorov: Sverkhprovodimost (KIAE) **3**, 1199-1213 (1990) [Engl. transl.: Superconductivity **3**, S50 (1990)]

6.58 A. Polyanskii, M.V. Indenbom, V.I. Nikitenko, Yu.A. Ossipyan, V.K. Vlasko-Vlasov: IEEE Trans. MAG-**26**, 1445-1447 (1990)

6.59 V.K. Vlasko-Vlasov, L.A. Dorosinskii, M.V. Indenbom, V.I. Nikitenko, A.A. Polyanskii, A.V. Antopnov, Yu.M. Gusev, G.A. Emelchenko: Sverkhprovodimost (KIAE) **4**, 1100-1109 (1991) [Engl. transl.: Superconductivity **4**, 1007-1016 (1991)]

6.60 K. Semba, H. Suzuki, A. Katsui, Y. Hidaka, M. Hikita, Sh. Tsurumi: Jpn. J. Appl. Phys. **26**, L1645-L1647 (1987)
6.61 V.K. Vlasko-Vlasov, M.V. Indenbom, Yu.A. Ossipyan: Pisma Zh. Eksp. Teor. Fiz. **47**, 312-315 (1988) [Engl. transl.: Sov. Phys.-JETP Lett. **47**, 375-378 (1988)]
D.E. Batova, V.K. Vlasko-Vlasov, V.A. Goncharov, G.A. Emel'chenko, M.V. Indenbom, Yu.A. Ossipyan: Zhn. Eksp. Teor. Fiz. **94**, 356-364 (1988) [Engl. transl.: Sov. Phys.-JETP **67**, 2376-2380 (1988)
6.62 H. Schmid, E. Burkhardt, E. Walker, W. Brixel, M. Clin, J.P. Rivera, J.L. Jorda, M. Francois, K. Yron: Z. Phys. B 72, 305-322 (1988)
6.63 Y. Isikawa, K. Mori, K. Kobayashi, K. Sato: Physica C **153-155**, 1471-1472 (1988)
6.64 T. Ishii, T. Yamada: Physica C **159**, 483-487 (1989)
6.65 A. Umezawa, G.W. Grabtree, K.G. Vandervoort, U. Welp, W.K. Kwok, J.Z. Liu: Physica C **162-164**, 733-734 (1989)
6.66 V.V. Moschchalkov, A.A. Zhukov, D.K. Petrov, V.I. Voronkova, V.K. Yanovskii: Physica C **166**, 185-190 (1990)
6.67 H. Adrian, W. Assmus, A. Höhr, J. Kowalewski, H. Spille, F. Steglich: Physica C **162-164**, 329-330 (1989)
6.68 A. Sulpice, P. Lejay, R. Tournier, J. Chaussy: Erophys. Lett. 7, 365-370 (1988)
6.69 A.D. Hibbs, A.M. Campbell: IEEE Trans. MAG-25, 2142-2145 (1989)
6.70 H. Küpfer, I. Apfelstedt, R. Flükiger, C. Keller, R. Meier-Hirmer, B. Runtsch, A. Turowski, U. Wiech, T. Wolf: Cryogenics 29, 268-280 (1989)
6.71 A. Dulĉic, R.H. Crepean, J.H. Freed, L.F. Schneemeyer, J.V. Waszczak: Phys. Rev. B **42**, 2155-2160 (1990)
6.72 C.P. Bean: Phys. Rev. Lett. **8**, 250 (1962); ibid. Rev. Mod. Phys. **36**, 31 (1964)
6.73 G. Van Tendeloo, D. Broddin, H.W. Zandbergen, S. Amelinckx: Physica C **167**, 627-639 (1990)
6.74 A. Polyanskii, L. Dorosinskii, M. Indenbom, V. Nikitenko, Yu. Ossipyan, V. Vlasko-Vlasov: J. Less Comm. Metals **164** & **165**, 1300-1307 (1990)
6.75 L.F. Schneemeyer, E.M. Gyorgy, J.V. Waszczak: Phys. Rev. B **36**, 8804-8806 (1987)
6.76 S. Senoussi, M. Oussena, G. Collin, I.A. Campbell: Phys. Rev. B **37**, 9792-9795 (1988)
6.77 M. Guillot, M. Potel, P. Gougeon, H. Noel, J.C. Levet, C. Choufeau, J.L. Tholence: Phys. Lett. A **127**, 363-365 (1988)
6.78 M. Kartsovnik, V.A. Larkin, V.V. Ryazanov, N.S. Sidorov, I.F. Shegolev: Pisam Zh. Eksp. Teor. Fiz. **47**, 595-597 (1988) [Engl. transl.: Sov. Phys.-JETP Lett. (1988)]
6.79 V.V. Moshchalkov, A.A. Zhukov, D.K. Petrov, V.I. Voronkova, V.K. Yanovskii: Physica C **166**, 185-190 (1990)
6.80 T.Y. Hsiang, D.K. Finnemore: Phys. Rev. B 22, 154-163 (1980)
6.81 M.M. Fang, D.K. Finnemore, D.E. Farrell, N.R. Bansal: Cryogenics **29**, 347-349 (1989)
6.82 M. Tuominen, A.M. Goldman, M.L. Mecartney: Phys. Rev. B **37**, 548-551 (1988); ibid. Physica C **153-155**, 324-325 (1988)
6.83 M. Tinkham: Phys. Rev. Lett. **61**, 1658-1661 (1988)
6.84 C.W. Hagen, R. Griessen: Phys. Rev. Lett. **62**, 2857-2860 (1989)
6.85 R. Griessen, J.G. Lensink, T.A.M. Schröder, B. Dam: Cryogenics **30**, 563-568 (1990)
6.86 M.V. Feigel'man, V.B. Geshkenbein, A.I. Larkin, V.M. Vinokur: Phys. Rev. Lett. **63**, 2303-2306 (1989)
6.87 D.E. Farell, J.P. Rice, D.M. Ginsberg, J.Z. Liu: Phys. Rev. Lett. **64**, 1573-1576 (1990)

6.88 L.N. Bulaevskii, V.L. Ginsburg, A.A. Sobyanin: Zh. Eksp. Teor. Fiz. **94**, 355-375 (1988) [Engl. transl.: Sov. Phys.-JETP (1988)]

6.89 M. Tuominen, A.M. Goldman, Y.Z. Chang, P.Z. Jiang: Phys. Rev. B **42**, 412-419 (1990)

6.90 D.E. Farell, S. Bonham, J. Foster, Y.C. Chang, P.Z. Jiang, K.G. Vandervoort, D.J. Lam, V.G. Kogan: Phys. Rev. Lett. **63**, 782-785 (1989)

6.91 Y. Iye, A. Watanabe, S. Nakamura, T. Tamegai, T. Terashima: Physica C **167**, 278-286 (1990)

6.92 M. Tachiki, S. Takahashi: Solid State Commun. **70**, 291-295 (1989); ibid. **72**, 1083-1086 (1989)

6.93 S. Iijima, T. Ichihashi, Yu. Shimakama, T. Manako, Y. Kubo: Jpn. J. Appl. Phys. **27**, L837-L840 (1988)

6.94 H.W. Zandbergen, G. Van Tendeloo, J. Van Landuyt, S. Amelinckx: Appl. Phys. A **46**, 233-239 (1988)

6.95 L. Fruchter, C. Aguillon, I.A. Campbell, B. Keszei: Phys. Rev. B **42**, 2627-2630 (1990)

6.96 J. van den Berg, C.J. van der Beek, P.H. Kes, J.A. Mydosh, M.J.V. Menken, A.A. Menovsky: Supercond. Sci. Technol. **1**, 249-253 (1989)

6.97 M. Nakao, K. Kawaguchi, H. Furukawa, K. Shikishi, Y. Matsuta: Physica C **162-164**, 677-678 (1989)

6.98 E.M. Gyorgy, R.B. van Dover, K.A. Jackson, L.F. Schneemeyer, J.V. Waszczak: Appl. Phys. Lett. **55**, 283-285 (1989)

6.99 F.M. Sauerzopf, H.P. Wiesinger, H.W. Wiesinger, H.W. Weber: Cryogenics **30**, 650-655 (1990)

6.100 U. Welp, W.K. Kwok, G.W. Grabtree, K.G. Vandervoort, J.Z. Lin: Appl. Phys. Lett. **57**, 84-86 (1990)

6.101 L. Dorosinskii, B. Farber, M. Indenbom, V. Nikitenko, A. Polyanskii, V. Vlasko-Vlasov: Ferroelectrics **111**, 321-331 (1990)
L.A. Dorosinskii, M.V. Indenbom, V.I. Nikitenko, Yu.A. Ossipyan, A.A. Polyanskii, V.K. Vlasko-Vlasov: In *Transport Properties of Superconductors, Progress in High Temperature Superconductivity*, ed. by R. Nikolsky (World Scientific, Singapore 1990) pp.166-170

6.102 D.L. Kaiser, F.W. Gayle, R.S. Roth, L.J. Swartzendruber: J. Mater. Res. **4**, 745-747 (1989)

6.103 J.P. Rice, D.M. Ginsberg: A method for producing untwinned $YBa_2Cu_3O_{7-\delta}$ crystals without subjecting them to stress. Preprint, University of Illinois (1990)

6.104 L.A. Dorosinskii, M.V. Indenbom, V.I. Nikitenko, B.Ya. Farber: Pisma Zh. Eksp. Teor. Fiz. **49**, 156-159 (1989) [Engl.. transl.: Sov. Phys.-JETP Lett. **49**, 182-187 (1989)]

6.105 L.T. Wille, A. Berera, D. de Fontaine: Phys. Rev. Lett. **60**, 1065-1068 (1988)

6.106 D. de Fontaine, M.E. Mann, G. Ceder: Phys. Rev. Lett. **63**, 1300-1303 (1989)

6.107 D. Agassi, R.V. Kasowski: In *Physics and Materials Science of High Temperature Superconductors*, ed. by R. Kossowsky, S. Methfessel, D. Wohleben, NATO ASI Series, Vol.E181 (Kluwer Academic, Dordrecht 1990) pp.547-558

6.108 M.A. Alario-Franco, C. Chailout, J.J. Capponi, J. Chenavas, M. Marezio: Physica C **156**, 455-460 (1988)

6.109 G. Van Tendeloo, H.W. Zandbergen, S. Amelinckx: Sol. State Commun. **63**, 389-393 (1987)

6.110 V.B. Geshkenbein: Zh. Exp. Teor. Fiz **94**, 368-373 (1988) [Engl. transl.: Sov. Phys.-JETP **67**, 2166 (1988)]

6.111 C. Keller, H. Küpfer, R. Meier-Hirmer, U. Wiech, V. Selvamanickam, K. Salama: Cryogenics **30**, 401-409 (1990)

6.112 W.K. Kwok, U. Welp, G.W. Grabtree, K.G. Vandervoort, R. Hulscher, J.Z. Liu: Phys. Rev. Lett. **64**, 966-969 (1990)
W.K. Kwok, U. Welp, K.G. Vandervoort, Y. Fang, G.W. Grabtree, J.Z. Liu: Appl. Phys. Lett. **57**, 1268-1270 (1990)
6.113 Y. Iye, S. Nakamura, T. Tamegai, T. Terashima, K. Yamamoto, Y. Bando: Physica C **166**, 62-70 (1990)
6.114 B. Roas, L. Schultz, G. Saemarn-Ischenko: Phys. Rev. Lett. **64**, 479-482 (1990)
6.115 R. Hergt, W. Andurä, K. Fisher, N.M. Tchebotaev, S.L. Town: Phys. Stat. Sol. (a) **119**, 241-250 (1990)
6.116 B. Janossy, R. Hergt, L. Fruchter: Physica C **170**, 22-28 (1990)
6.117 D. Shi, M.S. Boley, J.G. Hen, M. Tang, U. Welp, W.K. Kwok, B. Malecki: Supercond. Sci. Technol. **2**, 255-260 (1989)
6.118 F.J. Blunt, A.M. Campbell, P.P. Edvards, J.E. Evetts, P. Freeman, J. Johnson, J. Loram, K. Mirza, A. Putnis, E. Salje, W. Schmall: Physica C **162-164**, 1605-1606 (1989)
6.119 J.Z. Liu, M.D. Lan, P. Klavins, R.N. Sheton: Phys. Lett. A **144**, 265-268 (1990)
6.120 M.R. Koblischka, N. Moser, B. Gegenheimer, H. Kronmüller: Physica C **166**, 36-48 (1990)
6.121 L. Shimmele, H. Kronmüller, H. Teichler: Phys. Stat. Sol. B **147**, 361-372 (1988)
6.122 M. Tinkham: Helv. Phys. Acta **61**, 433 (1988)
6.123 D. Shi, M. Xu, M.S. Boley, U. Welp: Physica C **160**, 417-423 (1989)

# 7. Properties and Structure of Yttrium-Barium Cuprate Treated in Halogen Vapors

Yu. A. Ossipyan, O.V. Zharikov, and R.K. Nikolaev

The discovery of high-temperature superconductivity in the La-Ba(Sr)-Cu-O system stimulated an avalanche of works with the aim of searching for new compounds with high-$T_c$ values. As a result of these investigations, new complex superconducting oxides of perovskite structure were found, namely, Y-Ba-Cu-O, Ba-K-Bi-O, Bi(Ta)-Ba, (Sr)-Ca-Cu-O, Nd-Ce-Cu-O and their modifications, which exhibit different $T_c$ values. A simple analysis shows that at least two primary elements of the High-Temperaturte SuperConductor (HTSC) structure exist in each of these compounds. These are the cuprate planes of the $CuO_2$ type and a rather complicated sublattice of oxygen atoms [7.1,2]. Apparently, it is the structure of these elements that plays the decisive role in the charge redistribution and the formation of paired current carriers in these systems.

One of the main mysteries of high-temperature superconductivity is the unique role played by oxygen. This has motivated investigations into the real structure of the anion sublattice of HTSCs, for example, by studying the effects of substitution of oxygen by atoms of other oxidants. This approach was suggested by analogous experiments, e.g., from semiconductor physics, where investigations of the structure and properties of different elements or compounds subjected to direct doping and creation of defects of various types yielded fundamental results essential for understanding the physics of these phenomena and for practical application of these materials.

The pioneering work by *Ovshinsky* et al. [7.3] has given impetus to the investigation of the consequences of anion isomorphism in HTSCs. By using a method of solid-phase synthesis with partial substitution of a fraction of barium oxides by fluorides, they attempted to obtain a fluorine-containing ceramic Y-Ba-Cu-O and observed a high critical temperature $T_c \simeq 155$ K. Regrettably, other researchers in subsequent numerous works did not succeed in reproducing the above result and got an ambiguous answer for the occurrence of fluorine in the Y-Ba-Cu-O lattice, see, for instance [7.3-6]. In our opinion the main problem is that the synthesis of this multicomponent system is complicated and the conditions required for successful substitution of oxygen by halogen, are vague.

Later, an alternative approach was suggested for partial substitution in the system Y-Ba-Cu-O. In this approach, samples synthesized by standard procedures are treated in vapors of halogens or their compounds. In particular, *Ossipyan* et al. [7.7] reported the observation of a superconducting transition in initially dielectric samples of $YBa_2Cu_3O_6$, exposed to chlorine gas in the absence of air. We believe that, at least in the Y-Ba-Cu-O

Springer Series in Materials Science, Vol. 23
**The Real Structure of High-$T_c$ Superconductors**
Editor: V.Sh. Shekhtman 

system, this approach is reasonable and has advantages over the solid-phase synthesis procedure. This is related to the structural features of this compound and to the presence of a highly mobile oxygen atom in so-called *chain positions of the type* Cu1-O. This led to the idea to remove the seventh oxygen atom from the $YBa_2Cu_3O_7$ compound without cardinally violating the compound structure and then to attempt to fill the vacant positions with other oxidizing atoms.

In this chapter we restrict ourselves to considering the original experimental data concerned with investigation of the real structure and properties of Y-Ba-Cu-O samples exposed to halogen vapors, which is one of the prospective approaches to partial substitution of oxygen by halogen atoms in this system. (We shall not discuss the numerous publications containing extremely contradictory results related, mainly, to the investigation of the Y-Ba-Cu-O-F compound obtained by a solid-phase synthesis).

The structure of this chapter is as follows. First, we shall consider the sample synthesis, i.e., the procedure, specific conditions of treatment of the Y-Ba-Cu-O in vapors of halogens and their compounds. Data will be presented on fluorination, chlorination, bromination and iodination of ceramics, single crystals and films, along with some physical-chemical data, e.g. heats and rates of reaction. The superconducting properties and structure of the phases arising after such treatment are considered in detail in the two following sections. Then experiments with halogenated powders using NMR, NQR and Mössbauer techniques, quite powerful methods for investigation of the structure and properties of HTSCs, are presented. Finally, we shall briefly discuss the, still meager, results on the substitution O → X (X is a halogen) in other HTSC systems. Finally, we summarize the main results and propose possibilities for further studies. In principle, the entire chapter is devoted to the discussion of possible models, which explain to some degree the observed phenomena, and of the possible relationship between the real structure of halogenated compounds and the HTSC phenomenon.

## 7.1 The Halogenation Technique

The sample preparation involves treatment of preliminarily synthesized samples of Y-Ba-Cu-O with different amounts of oxygen in vapors of halogens or their compounds in the absence of air. The procedure is similar to all oxidants of the seventh group of the periodic table (F, Cl, Br, I).

### 7.1.1 Fluorination of Y-Ba-Cu-O Ceramics

Many publications on the doping of Y-Ba-Cu-O ceramic samples using a solid-gas reaction technique can be found. There are several groups of works in which different gas agents were employed for fluroination of Y-Ba-Cu-O ceramic samples, i.e., $F_2$ [7.4,5,8-11], $NH_4HF_2$ [7.12] and $NF_3$ [7.13-17].

The first group of works reported on the thermal treatment of $YBa_2Cu_3O_y$ (6.0<y<7.0) in a fluorine-gas flux. At a temperature above 450° C the compound decomposed with the formation of oxyfluorides. As suggested by X-ray data, thermogravitometric and chemical analyses, at T < 450° C fluorine is incorporated into the sample. After fluorination the amount of oxygen in the sample is decreased. For example, it was pointed out that a fluorinated $YBa_2Cu_3O_{6.885}$ sample acquired the composition $YBa_2Cu_3O_{6.465}F_{0.74}$ [7.4]. These samples exhibited a superconducting state transition at $T_c \simeq$ 80-90K.

However, *Stuart* et al. [7.8] reported that fluorination is always followed by a partial decomposition of the sample, giving rise to multiphase reaction products. Therefore dilution of fluorine by an inert gas was tested as a way to optimize the annealing conditions. In [7.8] $YBa_2Cu_3O_6$ was annealed in diluted fluorine (10% $F_2$ in $N_2$) at 165° C for 6h. An analogous fluorination procedure was used in [7.5, 9, 10] but no evidence for fluorine incorpration into the lattice of Y-Ba-Cu-O was obtained. In [7.11] fluorine concentration has been reported to be only in a $\simeq 1\,\mu$m-thick grain surface.

The experiments in which $NF_3$ is used as a soft fluorinating agent are worth noting. $NF_3$ allows one to avoid both problems arising with $F_2$: the high activity of pure fluorine which leads to a decomposition of Y-Ba-Cu-O, and the disadvantages of dilution or temperature decrease, which prevents successful incorporation of the F atoms into the lattice.

*LaGraff* et al. [7.14] treated $YBa_2Cu_3O_{7-\delta}$ in a flux of pure $NF_3$ gas at 300° C for 5÷300 min. Fluorination of the tetragonal phase led to decomposition of the compound. This is possibly due to the high chemical activity of fluorine with respect to this phase. After treatment of the oxygen-deficient phase ($\delta$ = 0.35) a monotonic dependence of the lattice parameters on the fluorine content in the sample was observed, whereas no impurities were found by X-ray methods. In all the cases $T_c$ was 90 K.

A detailed study of fluroination processes of Y-Ba-Cu-O ceramics using $NF_3$ was conducted by *Perrin* et al. [7.13, 15-17]. They proposed to ease the fluorination conditions by employing diluted $NF_3$ (4% $NF_3$ in $N_2$). They determined the optimal conditions of fluorination (T = 300° C) and proved that fluorine does incorporate into the Y-Ba-Cu-O lattice (by X-ray, NMR and other methods, see below). The sample weight grew as a function of the annealing time, the gain being greater for samples with lower initial oxygen content (samples with $O_7$, $O_{6.7}$ and $O_6$ were studied). All the $NF_3$-treated samples demonstrated diamagnetic response.

### 7.1.2 Chlorination of Y-Ba-Cu-O Ceramics and Films

Like in the case of fluorination, various gaseous agents of different chemical activity with respect to the solid phase, i.e., pure $Cl_2$ [7.7, 18-23], $CCl_4$ [7.22], and $PCl_5$ [7.20] gases, were employed to chlorinate Y-Ba-Cu-O samples.

Exposure of tetragonal $YBa_2Cu_3O_6$ samples to chlorine gas can give rise to the orthorhombic superconducting phase ($T_c \simeq$ 90K) only when thermally treated in the narrow temperature range 160÷180° C for several minutes. (At least one work, however, [7.19] reports a thermal treatment temperature T $\simeq$ 120° C). Usually, chlorinated samples are not single-phase and contain significant amounts of the starting phase. When the thermal treatment temperature is increased above the optimum one, the sample disintegrates with the formation of auxiliary reaction products. At a too low temperature the reaction either does not proceed at all or proceeds very slowly. An increase in the amount of oxygen leads to an increase in the temperature for an optimum thermal treatment. The chlorination process has a complicated kinetics and reaches to saturation [7.23]. In their preliminary report *Chu* et al. [7.24] showed the possibility of chlorination of thin films.

*Perrin* et al. [7.22] employed $CCl_4$ gas to generate mild chlorination conditions. They studied the chlorination process for samples with different amounts of oxygen: $O_6$, $O_{6.7}$ and $O_7$. The optimum treatment temperature (250÷270° C) was determined and the possibility of obtaining practically single-phase orthorhombic superconducting samples was demonstrated. The kinetics (as measured by the rate of weight gain by the samples) depended on the initial oxygen content.

For the oxygen → chlorine substitution *Klimenko* et al. [7.20] proposed to employ the reaction between the oxygen-saturated phase of $YBa_2Cu_3O_7$ and $PCl_5$. Upon heating, $PCl_5$ dissociates to $PCl_3$ + $Cl_2$ and, as the oxygen is substituted by chlorine, the released oxygen atoms bind to $PCl_3$ to form $OPCl_3$. They used $^{31}$P NMR measurements to register the appearance of $OPCl_3$ and confirm the O ⟷ Cl substitution. The obtained superconducting chlorine-containing samples were single-phase and had the chemical composition $YBa_2Cu_3O_yCl_x$ (x = 0.6, 1.57 and 3.12 in different samples). However, we note that because of the similarity in the $T_c$ and crystal lattice parameters in the initial and chlorinated samples, the O → Cl substitution is not proven.

### 7.1.3 Bromination and Iodination of Y-Ba-Cu-O Ceramics and Single Crystals

Y-Ba-Cu-O samples are brominated and iodinated by exposing them to pure $Br_2$ and $I_2$ gases at different temperatures and pressures in the absence of air. Iodination of Y-Ba-Cu-O ceramics was considered in [7.18, 19, 21, 25-27], and of single crystals in [7.27]. Bromination of ceramic samples was studied in [7.18, 19, 21, 23, 25, 28], of single crystals in [7.23.27] and of films [7.28].

The optimum (with regard to the characteristics of the final product) temperature of bromination varies from 180 to 260° C, and the best bromine vapor pressure from 1 to 3 atm; a further pressure increase (as shown by our data) leads to the appearance of additional reaction products and a decrease in the fraction of the superconducting phase. The time period of the process depends on the temperature of the experiment and is normally not large (1 h). An increase of the bromination temperature above the optimal one, like for other halogenations, is followed by the appearance of decomposition products, supposedly metal bromides and oxybromides. The chemical activity of Br is much lower than that of Cl and F, which makes it possible, in a number of cases, to obtain single-phase reaction products with very high fractions of superconducting phase, even after treatment of a tetragonal phase with $O_{6.1}$-$O_{6.2}$ [7.23,28]. However, other studies indicated the presence of the starting tetragonal phase in the brominated compound [7.18,19,25].

*Radouski* et al. [7.23] and *Ossipyan* et al. [7.27] carried out the bromination of initially tetragonal single-crystal samples. They showed that the characteristic time of halogenation in this case increases appreciable (to $\simeq$ 24h), however, the fraction of the orthorhombic superconducting phase never reaches 100% [7.23]. The possibility of bromination of thin dielectric tetragonal Y-Ba-Cu-O films was indicated in [7.28].

The process of iodination of Y-Ba-Cu-O samples is similar to bromination. However, it requires higher temperatures (200° -500° C), with a temperature of $\simeq$300° C being optimal, and the treatment duration is increased to hours or even days.

The structure and properties of the iodinated reaction product seem to have a complicated dependence on the iodine pressure. From our data a range of 1÷7 atm is best, a further pressure increase gives rise to impurities. Under the best conditions, the iodinated orthorhombic phase is single-phase and does not contain impurities [7.26,29].

Single crystals of Y-Ba-Cu-O-I have been prepared using a similar procedure [7.18,27]. Sections 7.2,3 present a detailed description of the structure and superconducting properties of iodinated crystals. It should be noted that these samples are non-single-phase.

## 7.2 Superconducting Properties of Y-Ba-Cu-O Treated in Halogen Vapors

Figure 7.1 depicts a typical example of the temperature dependence of the magnetic susceptibility $\chi_{ac}(T)$ (magnetic screening) of initially dielectric ceramic samples of $YBa_2Cu_3O_{6.1}$ treated in vapors of I, Br and Cl. The onset of the superconducting transition $T_{co}$ is, respectively, 50÷55 K, 75÷80 K, and 90÷92 K in these samples, and the transition proceeds throughout the entire temperature range down to liquid-helium temperatures. The transition curves of some brominated samples exhibit steps at T $\simeq$ 50÷60 K. Further experiments enabled us to select the optimal regimes of thermal treatment for iodination. After a short preliminary thermal treatment at 350° -400° C for $\simeq$ 0.5 h, the samples were exposed to iodine vapors at 300° C for 8÷12 h. This led to noticeable narrowing of the superconducting transition width to $\Delta T_c \simeq$ 10÷12 K in the best samples. The magnitude of the screening signal was normally 1-10% for chlorinated and 10÷100% for

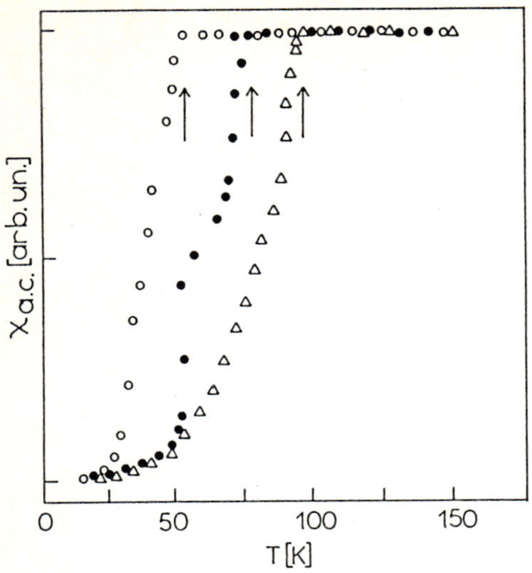


**Fig.7.1.** Temperature dependence of the magnetic susceptibility $\chi_{ac}$ of chlorinated (Δ), brominated (•) and iodinated (o) samples of the $YBa_2Cu_3O_{6.1}$ ceramics

brominated and iodinated samples of that of a specially prepared ceramic sample of $YBa_2Cu_3O_{6.9}$ of similar form and size.

There is a certain spread in the values of critical temperature obtained by various researchers: for chlorinated samples $T_c$ = 78 K [7.19], 90÷93 K [7.20], 85-90 K [7.22], 90 K [7.23]; for brominated samples $T_c$ is 90 K [7.19,23], 92 K [7.78]; and for iodinated samples 60 K [7.19]. In the numerous works on fluorinated Y-Ba-Cu-O the critical temperature was $T_{co} \simeq 90$ K in those cases when the researchers succeeded in obtaining the superconducting phase.

Measurement of the Meissner effect in halogenated samples yielded unusual results. It is known that even in the best oxygen-saturated ceramic samples of $YBa_2Cu_3O_{6.9}$ the $M(T) = M_{FC}(T)/M_{ZFC}(T)$ amplitude ratio [$M_{FC}(T)$ being the Meissner effect magnitude and $M_{ZFC}(T)$ the magnetic screening signal used to determine the amount of the superconducting phase] does not normally exceed 30÷50%. Without going into the meaning of M(T) < 100%, which is connected with flux pinning and the real field distribution in the sample, we note that in halogenated samples M(T) is very large. As an example, Fig.7.2 exhibits the data for magnetic measurements for an iodinated ceramic sample. The M value is seen to reach 90÷95%. Other researchers also noted large M(T) values in iodinated [7.19], brominated [7.28] and chlorinated [7.23] ceramic Y-Ba-Cu-O samples. (The results of M(T) measurements in fluorinated samples are inconsistent [7.14, 15]). This manifests the presence of a considerable amount of superconducting phase in halogenated samples. However, we must assume that using the M(T) value as a quantitative criterion is not quite correct since it does not take into account such factors as the distribution of the magnetic flux in a multiconnected system of superconducting and normal particles (grains) in a sample, flux pinning, etc. Nevertheless, this phenomenon needs study.

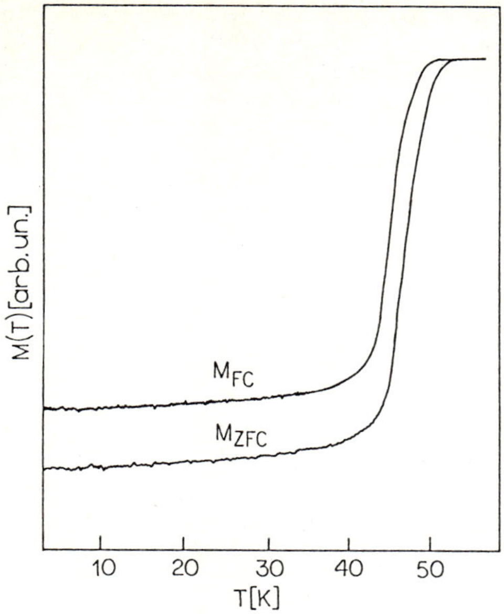


**Fig.7.2.** Temperature dependences of the magnetic moment M(T) of an iodinated $YBa_2Cu_3O_{6.1}$ sample ceramic ($M_{FC}$: magnitude of Meissner effect, $M_{ZFC}$: magnetic screening)

Halogenation of compact ceramic Y-Ba-Cu-O samples increases their tendency to crumble. In our experiments, for example, chlorinated samples always turned to powder. For this reason no data on the conductivity of halogenated samples can be obtained. We did succeed in measuring the temperature dependence of the resistance R(T) in a number of brominated samples. Figure 7.3 presents an example of such measurements. The curve exhibits portions of strong decreases at $T \simeq 90$ K and $T \simeq 50 \div 60$ K which may be interpreted as the superconducting transitions of two phases with

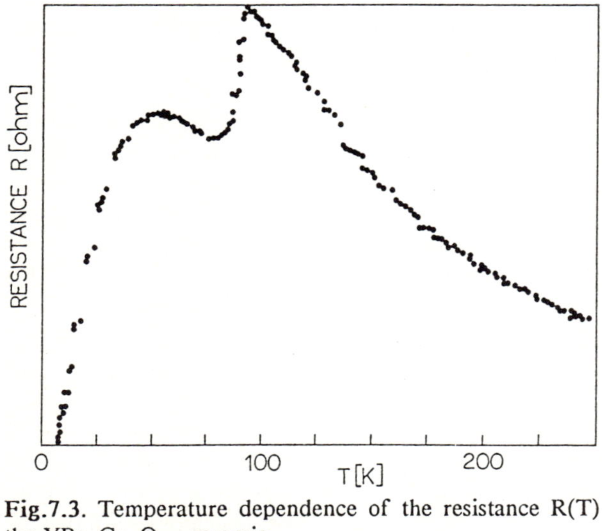


**Fig.7.3.** Temperature dependence of the resistance R(T) of a brominated sample of the $YBa_2Cu_3O_{6.1}$ ceramic

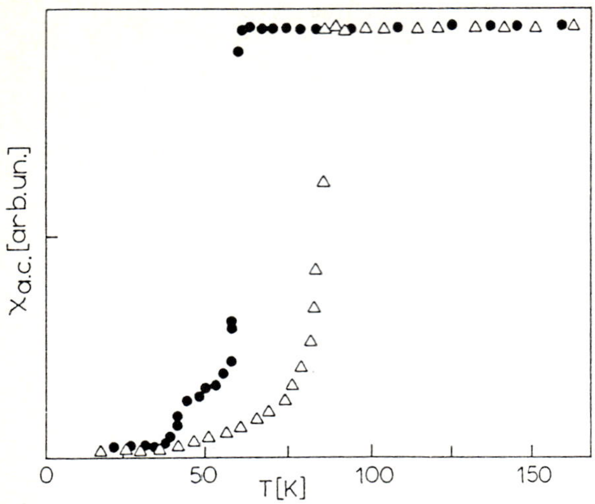


**Fig.7.4.** Temperature dependence of the magnetic susceptibility $\chi_{ac}(T)$ (the modulating field is perpendicular to the ab-plane): (•) single crystal of $YBa_2Cu_3O_{6.1}I_x$, (Δ) single crystal of $YBa_2Cu_3O_{6.1}Br_x$

different $T_c$. In the normal state R(T) has the behavior characteristic of a semiconductor. We believe that this is determined by the charge-carrier fluxes in the system through regions with different electric properties. In iodinated samples, probably due to poor connections between conducting regions, the resistance was very high and, because of the background due to semiconductor behavior, no superconducting transition was observed.

To conclude this section we shall describe the electric and magnetic properties of monocrystalline iodinated [7.27] and brominated [7.23,27] samples of Y-Ba-Cu-O. Figure 7.4 displays the data of a $x_{ac}(T)$ measurement in initially dielectric tetragonal single crystals of Y-Ba-Cu-O treated in iodine and bromine vapors. A sharp transition to the superconducting state with $T_{co} \simeq 58$ K is observed in the iodinated and $T_{co} \simeq 85$ K brominated crystals, in this case 80÷85% of the change occurs in a narrow temperature interval $\Delta T_c \simeq 2 \div 3$ K. Figure 7.5 shows the measurement of the magnetic moment M(T) (Meissner and screening effects) in an iodinated Y-Ba-Cu-O single crystal for two orientations of the external magnetic field H with respect to the *ab*-plane. When **H**⊥ *ab*, $M_{ZFC}(T)$ changes drastically and the transition width agrees with the data obtained from $x_{ac}(T)$ measurements in the case **H**|| *ab* the transition is extended down to 4.2 K. This may be explained, e.g., if the inequality $H_{cl}^{\parallel} < H < H_{cl}^{\perp}$ holds, where $H_{cl}^{\parallel}$ and $H_{cl}^{\perp}$ are the lower critical fields corresponding to two crystallographic orientations with respect to the *ab*-planes. Another characteristic feature of these curves is the considerable flux trapping [a small change of $M_{FC}(T)$] which probably reflects a strong pinning of vortices in the iodinated crystal.

The magnetic moment M for the case **H**|| *ab* appeared to be dependent on the value of the external magnetic field for all **H** values (and for all temperatures). Estimation of the lower critical field (or more exactly the

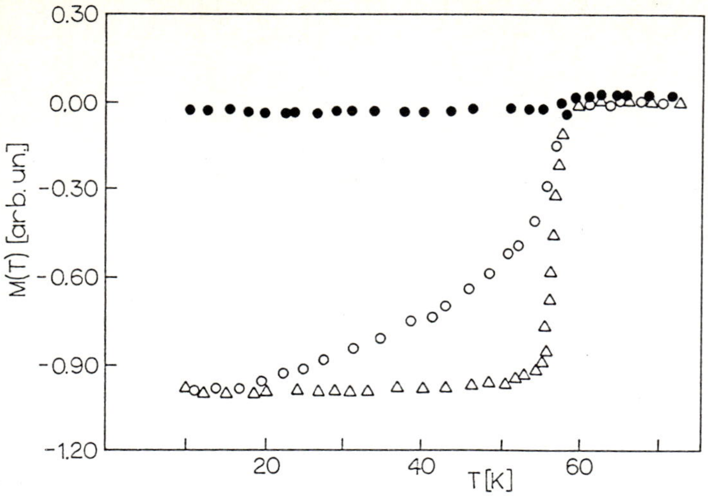


**Fig.7.5.** Temperature dependence of the magnetic moment M(T) of an iodinated single crystal, external field H = 1 Oe. (•) magnitude of Meissner effect $M_{FC}$, **H**⊥ *ab*-plane, when **H**|| ***ab*** the $M_{FC}$(T) curve is practically coincident with this one), (Δ) magnetic screening $M_{ZFC}$, **H**⊥ ***ab***, (○) $M_{ZFC}$, **H**|| ***ab***

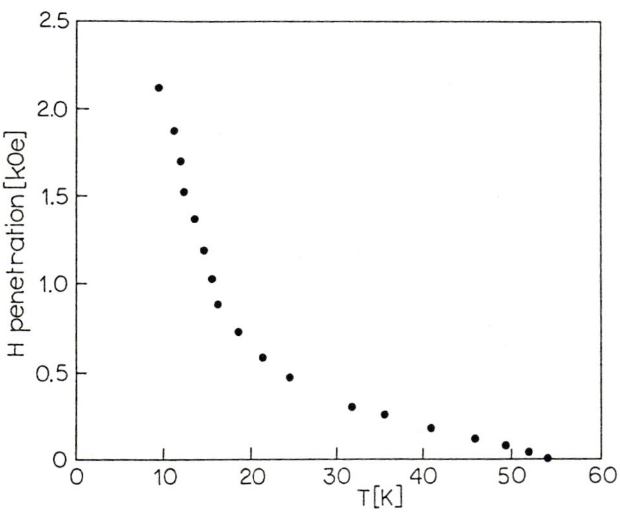


**Fig.7.6.** Temperature dependence of the penetration field $H_*^{\perp}$(T) of iodinated single crystals

penetration field $H_*$) at 10 K yields $H_*^{||} \le 2$Oe. When **H**⊥ ***ab*** one can correctly determine the $H_*$ value by using linear extrapolation of M/M(T) to -1. The temperature dependence of the field $H_*^{\perp}$(T) of an iodinated single crystal is demonstrated in Fig.7.6. Note the anomalous behavior of $H_*^{\perp}$(T) that is encountered, however in layered and high-$T_c$ superconductors. Our

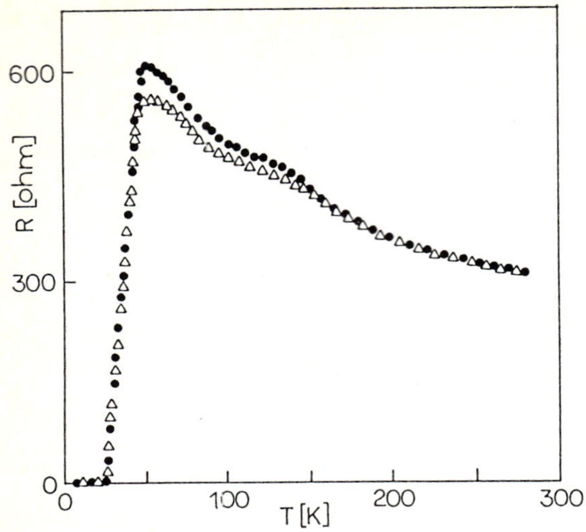


**Fig.7.7.** Temperature dependence of the resistance R(T) of an iodinated single crystal (the current j is parallel to the *ab*-plane). (Δ) j = 1 μA, (•) j = 10 μA

data yields a rough estimate of the anisotropy of the lower critical field in single crystals: $H_{cl}^{\perp}/H_{cl}^{\parallel} \simeq H_*/H_*^{\parallel} \simeq 10^2$.

Figure 7.7 depicts the temperature dependence of the resistance R(T) of an iodinated single crystal when the measuring current is parallel to the *ab*-plane. The temperature dependence has semiconductor character, the R(T) value increases by approximately a factor of two with decreasing temperature from 300 K to 70 K. Regrettably, we did not succeed in measuring R in the perpendicular direction and estimating the anisotropy of the conductance. This is because it is impossible to determine the real thickness of the conducting crystal layer because of the complicated current distribution in an iodinated single crystal with inhomogeneous structure (Sect.7.4). The value R(T) = 0 in this case is reached at T ≃ 35 K, and the upper estimate for the resistivity yields $R_{300K} \simeq 1\ \Omega\cdot$cm.

Thus, these studies on the superconducting properties of ceramic samples, single crystals and films of Y-Ba-Cu-O treated in halogen vapors, support convincingly the recovery of superconductivity in such samples. The appearance in, of at least, two types of superconducting phases with $T_c$ values in the proximity of 90 K and 60 K, in different samples and after different treatments of of particular significance. Apparently this is linked to the two orthorhomibc phases 1 and 2, in Y-Ba-Cu-O compounds which differ in the concentration of oxygen atoms ($O_{7.0}$ and $O_{6.5}$) as well as the type of oxygen ordering in chains and the concentration of the carriers (holes) [7.30-32]. This suggests that halogen incorporation in these systems (when it doesn't lead to decomposition) leads to the formation of one (or two at a time) of the possible stable states, typical of Y-Ba-Cu-O.

## 7.3 Structural Features of Halogenated Phases

Structural investigations of samples treated in halogen vapors are very important for the understanding crystallophysical and chemical features of the high-$T_c$ superconducting phenomenon. In addition, they are interesting because they involve the analysis of the structure of new phases that may contain atoms forced into these phases. Unfortunately, the data on structural studies are contradictory and do not allow final conclusions about the arrangement of halogen atoms in the Y-Ba-Cu-O lattice. This circumstance reflects the complexity of these systems and the insufficient quality of the samples employed. As will be shown below, different situations may be realized for different halogens due to differences in the electronic structure of atoms, chemical activity, ion size, etc.

A feature noted by practically all researchers is the formation of an orthorhombic Y-Ba-Cu-O phase in initially tetragonal samples after they are treated in vapors of different halogens (F, Cl, Br, I), with crystal lattice parameters close to those of oxygen-rich ceramics [7.6, 14, 19, 25, 26, 28]. However, a detailed study of the structure, e.g., of iodinated samples, demonstrates certain variations [7.27]. Figure 7.8 gives the values of the crystal lattice parameters ***a***, ***b***, ***c*** in iodinated ceramics versus a, b, c values as a function of $O_y$ in Y-Ba-Cu-O. The set of ***a***, ***b*** and ***c*** values in Y-Ba-Cu-O-I is seen to have nonnegligible differences outside the error limits of the measurements from, e.g., the oxygen orthorhombic-2 phase with the same ***c*** value. It is important to note that in these experiments the same ceramic sample was used both for oxidation and iodination.

The high resolution of the diffraction patter, with narrow peaks, and the absence (within the detection limits) of impurity phases at least in the best iodinated samples, allows unambiguous conclusions to be drawn (Fig.

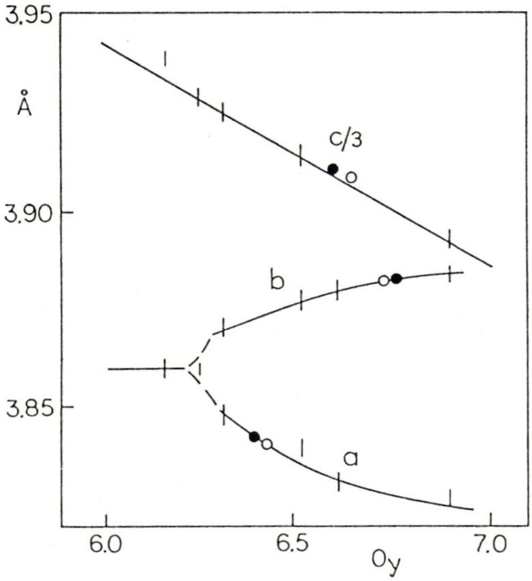


**Fig.7.8.** Comparison of the a,b and c lattice parameters of iodinated single crystal and ceramic samples with the a,b and c dependence on the oxygen content $O_y$ in the $YBa_2Cu_3O_y$ system. (|) ceramic samples of $YBa_2Cu_3O_y$ (our data), (•) iodinated ceramic sample, (○) iodinated single crystal

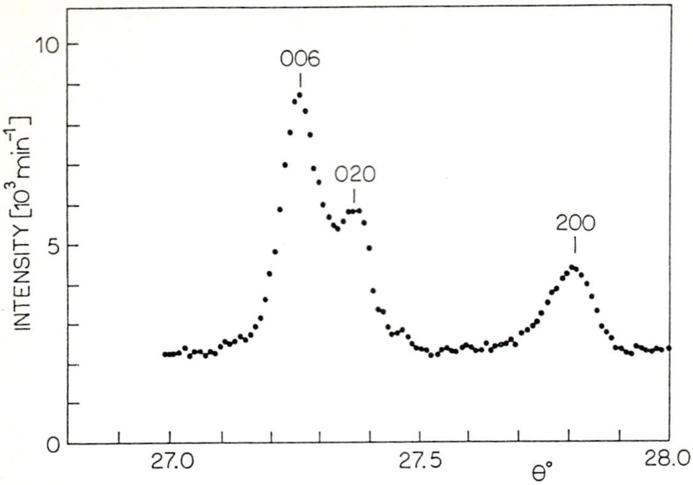


**Fig.7.9.** Fragment of X-ray powder diffraction pattern of a typical I-doped sample; $CoK_{\alpha 1}$ radiation

7.9). Note that the main impurity phase with respect to the orthorhombic phase is the starting tetragonal phase. The intensity of this peak reflects the completeness of the halogenation process and depends on the technique and conditions of the treatment. Thus, bromination and also chlorination and fluorination in harsh conditions leads to partial amorphization (destruction of the crystalline structure) and retaining of the starting tetragonal phase, which coexists with the arising orthorhombic, near-orthorhombic-1, phase [7.7,25,26]. Milder halogenation conditions make it possible to obtain practically single-phase brominated [7.28], chlorinated [7.20,22] and fluorinated samples [7.14,33,34]. These works also indicate a similarity of the lattice parameters with the values characteristic of the normal oxygen orthorhombic-1 phase. Chlorination of orthorhombic $YBa_2Cu_3O_{6.9}$ samples by $PCl_5$ [7.20] and fluorination of $YBa_2Cu_3O_y$ with different **y** [7.33,34] also leads to the formation of the orthorhombic-1 phase. Thus taking into consideration only the best-quality single-phase samples, one may conclude that the structure of Y-Ba-Cu-O treated in F, Cl, or Br is close to the orthorhombic-1 phase, and that of iodinated samples to the orthorhombic-2 phase.

This suggests two alternatives for the consequences of Y-Ba-Cu-O halogenation. The first involves oxygen-halogen substitution and incorporation of the latter into well-defined positions in the crystal lattice. In the second, oxygen atoms occur in their usual position of Cu(1)-O type due to the chemical interaction of Y-Ba-Cu-O ceramics with the halogen and, possibly, its partial decomposition. Therefore it is important to confirm the presence and positions of halogen atoms in the crystal lattice. The type and position of atoms in a crystal lattice can usually be found by a Rietweld analysis of the experimental diffraction patterns, using different programs of profile analysis. Unfortunatley, analytic data yield different, often contradictory results. For example, *Radousky* et al. [7.28] failed to obtain any evidence for the occupation of vacant O4 or O5 positions by bromine atoms because the occupation value did not exceed the experimental error by more than several percent. A neutron-diffraction pattern analysis of fluorinated ceramics by *LaGraff* et al. [7.14] and *Perrin* et al. [7.33] concluded that fluorine atoms occupy vacant oxygen positions in the Cu(1) phase. An unexpected result was obtained for chlorinated ceramics [7.35]. There, a two-phase sample showed substitution of oxygen by chlorine in the O(2) and O(3) positions in the tetragonal phase (!) whereas the orthorhombic phase exhibited the presence of only oxygen atoms.

Application of other methods in structural studies of halogenated samples is also of great interest. Such works are scarce, however, and do not give an unambiguous answer about the arrangement on halogen atoms in the lattice. Methods of anomalous scattering and EXAFS have been employed to investigate brominated ceramics [7.36]. These researchers claimed that bromine incorporates into the vacant positions of a Cu(1) plane. However, Raman spectra of iodinated [7.37] and brominated [7.28] samples compared with the spectra of $YBa_2Cu_3O_y$ show in our opinion, that the oxygen is in chain positions Cu(1)-O.

Thus, structural studies of ceramic samples prove significant structure modifications resulting from halogenation. However, they do not allow conclusions about the arrangement of halogen atoms in the Y-Ba-Cu-O lattice to be drawn nor about the transition of oxygen to the O(4) or O(5) positions due to partial decomposition of the sample and formation of metal halogenides or oxyhalogenides. Naturally, a more complicated process is not excluded. In addition, various structures may arise upon treatment by different halogens (F, Cl, Br, I) owing to the difference in their properties.

To conclude this section, we shall briefly discuss the data obtained for iodinated single crystals. In [7.27] angular scanning topography and precision diffractometry were employed to study the structure of these species. Details of topograms and diffraction patterns can be found in Sect.3.5. The following lattice parameters were obtained: $\boldsymbol{a} = \boldsymbol{b}$ = 3.855 Å, $\boldsymbol{c}$ = 11.822 Å for the tetragonal matrix, $\boldsymbol{a}$ = 3,836 Å, $\boldsymbol{b}$ = 3.877 Å, $\boldsymbol{c}$ = 11,722 Å for iodinated layers of the orthorhombic phase. From the dependences $\boldsymbol{a}$, $\boldsymbol{b}$, $\boldsymbol{c}$ on the oxygen content $O_y$, one may obtain c = 11÷722 Å which corresponds to a hypothetical oxygen content of y = 6.57 (Fig.7.8 and [7.27, 30]). The value of $\boldsymbol{a}$ = 3.886 Å corresponds to y = 6.57, and $\boldsymbol{b}$ = 3.877 Å to y = 6.25. The inconsistency of these parameters with the known concentration dependences of $\boldsymbol{a}$, $\boldsymbol{b}$ and $\boldsymbol{c}$ on oxygen content is indirect evidence for the presence of iodine in the lattice. Also (for ceramic samples) it directly proves that the systems Y-Ba-Cu-O and Y-Ba-Cu-O-I are different.

## 7.4 Investigations of Atomic Nuclei in Halogen Substituted Ceramics

### 7.4.1 Nuclear Quadrupole Resonance and Nuclear Relaxation of $^{63}$Cu in the Y-Ba-Cu-O-I Ceramic

A single-phase polycrystalline powder with a mean particle size of <50 $\mu$m in a paraffin matrix was studied in a pulsed NQR spectrometer using spin echo signal accumulation [7.38, 39]. Several sets of samples, prepared from the same ceramics, were used. The starting materials were orthorhombic samples of $YBa_2Cu_3O_{6.9}$, synthesized by standard techniques, with $T_c$ = 90 K. These samples were then treated in argon at different temperatures, which yielded tetragonal dielectric samples of $YBa_2Cu_3O_6$, and superconducting orthorhombic $YBa_2Cu_3O_{6.56}$ with $T_c$ = 55÷60 K. The oxygen content was determined by a standard iodimetric titration method and controlled by using the dependence of the crystal lattice parameters ***a***, ***b*** and ***c*** on $O_y$ (Fig.7.8). Finally, the tetragonal samples were treated in iodine by the technique described in Sect.7.1.3. The choice of the samples with oxygen content y = 6.56 was not arbitrary: these samples, like the iodinated ones, have similar values of $T_c$.

The $^{63,65}$Cu NQR spectra are shown in Fig.7.10. The spectrum of the starting orthorhombic sample (Fig.7.10a) agrees with the published data, i.e., a 31.5 MHz resonance line assigned to Cu2 atoms, and a 22.9 MHz line to Cu1 atoms [7.40]. The spectrum of the tetragonal $YBa_2Cu_3O_{6.1}$ sample (Fig.7.10c is also consistent with the numerous results available). Namely, two spectrum lines at 30.1 and 27 MHz correspond to the NQR signals of $^{63}$Cu and $^{65}$Cu isotopes, respectively, in the chain positions Cu1-O. Antiferrogmagnetic ordering of the $Cu^{2+}$2 ions in copper-oxygen planes gives rise to a strong magnetic field in this case, therefore no NQR signal is observed in the region of 31.5 MHz. However, a quadrupole-perturbed NMR spectrum in the internal magnetic field at frequencies 70÷110 MHz does arise [7.40, 41]. In the oxygen-deficient orthorhombic sample of $YBa_2 \cdot Cu_3O_{6.5}$ the resonance lines have frequencies 31.5 and 29.1 MHz. Antiferromagnetic order disappears in this compound, superconductivity appears, and the NQR signals are attributed to copper atoms in the Cu2 positions.

The NQR spectrum of the iodinated sample, as shown in Fig.7.10d, is very much like that of an oxygen-deficient sample. The quadrupole spin echo signals are observed at 31.4 and 29.1 MHz. It should be noted, however, that the intensity of the NQR resonance lines at 31.4 MHz is almost an order of magnitude higher than that of $YBa_2Cu_3O_{6.56}$ at 31.5 MHz. The resonance line width of an iodinated sample is $\Delta\nu_Q$ = 200 kHz (for comparison, in $YBa_2Cu_3O_{6.9}$ $\Delta\nu_Q$ = 700KHz), which is comparable to the data for highest-quality samples of $YBa_2Cu_3O_{6.9}$.

Another feature of the NQR spectrum of the iodinated sample is the appearance of a weak signal at 23.9 MHz. However, we did not observe a line in the region of 22 MHz, characteristic for the Cu1 crystallographic positions in the orthorhombic phase. At present it is difficult to correctly

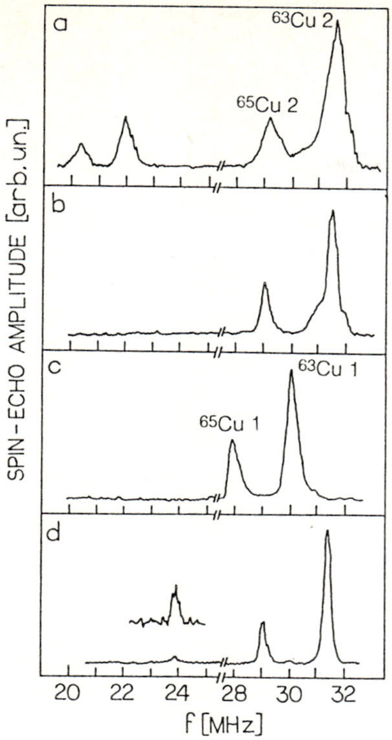


**Fig.7.10.** $^{63,65}$Cu NQR spectra at T = 4.2 K. **(a)** $YBa_2Cu_3O_{6.9}$, **(b)** $YBa_2Cu_3O_{6.56}$, **(c)** $YBa_2Cu_3O_{6.1}$, and **(d)** $YBa_2Cu_3O_{6.1}I_x$

interpret this feature since the comparison of these data with those for an oxygen-deficient sample is impossible, due to the absence of a noticeable signal in this spectral region in $YBa_2Cu_3O_{6.5}$.

Comparing the four $^{63,65}$Cu NQR spectra suggests that the intense resonance lines occurring after iodination are possible due to copper atoms in the Cu2 positions. In this case the spectrum directly implies that the $CuO_2$ planes in the iodinated ceramics are not distorted and the small width of the observed lines indicates a sufficiently high degree of ordering of the crystal lattice. This correlates with the narrowness of the diffraction peaks on the X-ray photographs of iodinated samples (Fig.7.9). The appearance of weak NQR signals in the region of 23.9 MHz should be attributed to Cu1 chain atoms. The spectra do not show whether iodine is incorprated into vacant oxygen positions in the Cu1-O plane. However the structure and electronic state must be substantially reconstructed after iodination. The high degree of ordering of $CuO_2$ planes in iodinated samples should be noted.

If we assume that iodine is localized in vacant positions in Cu1-O planes, then the NQR data allow one to estimate the hypothetical charge state of iodine in the crystal lattice. Such a calculation based on the electric field gradient on copper cores in the Cu1 positions has been given in [7.39], and a small value (-0.5) of negative charge on iodine atoms is derived. This is close to the Mössbauer estimate (see below).

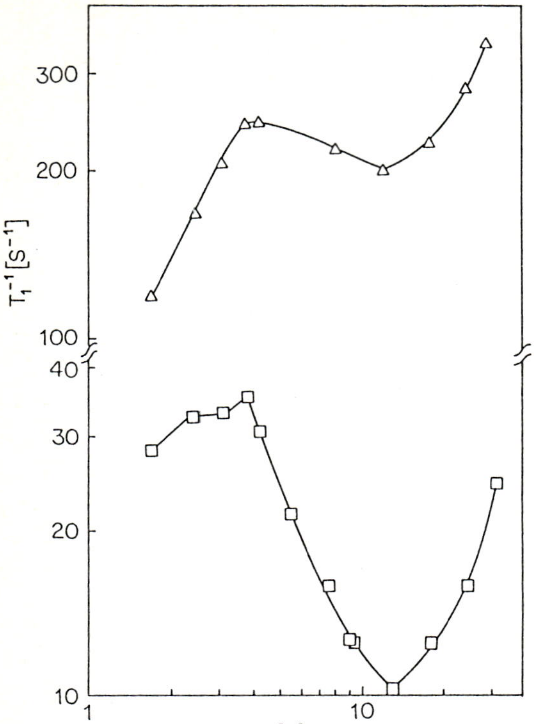


**Fig.7.11.** Temperature dependence of the spin-lattice relaxation rate $1/T_1(T)$ in ($\Delta$) $YBa_2Cu_3O_{6.1}$ and ($\square$) Y-B-Cu-O-I

The temperature dependence of the spin-lattice relaxation rate $T_1^{-1}(T)$ in iodinated and tetragonal samples was measured by $^{63}$Cu NQR methods in the range of 1.7÷30 K. [The recovery of the equilibrium value of nuclear magnetization in the samples had an exponential time dependence, $A(t) = A_0[1-\exp(t/T_1)]$. The experimental data are displayed in Fig.7.11. The dependence is nonmonotonic, which was also previously observed [7.42,43]. At present the nature of such anomalous behavior in unclear. A possible cause may be the existence of a magnetic phase transition in this temperature range.

The relaxation rate in the iodinated sample is markedly different both from the results obtained for a tetragonal sample and the available data for the orthorhombic phases of $YBa_2Cu_3O_7$ and $YBa_2Cu_3O_{6.5}$ [7.40]. In an iodinated sample the spin-lattice relaxation is rapid, at 4.2 K it reaches $1/T_1 \simeq 30 \div 40\ s^{-1}$. (Note that the small intensity of the NQR signals at 23.9 MHz in the iodinated sample complicates accurate determination of the relaxation time. A rough estimation, however, yields $1/T_1 \simeq 1 s^{-1}$). Thus the difference in the relaxation parameters again supports the assignment of the resonance lines in the iodinated sample to copper atoms in Cu2 positions and indicates cardinal changes in the electron spin system.

### 7.4.2 NMR Study of Fluorinated Samples

Y-Ba-Cu-O samples treated by dilute $NF_3$, with different starting oxygen contents (y = 6.0, 6.7 and 7) were studied by NMR [7.15]. In the three halogenated samples the fluorine content was x = 1.4, 0.7, 0.56. It is important to note that the critical temperature was $T_c = 90$ K for x + y = 7 and $T_c \leq$ 50 K in $YBa_2Cu_3O_6F_x$, whereas the lattice symmetry did not change from that of Y-Ba-Cu-O. The $^{19}F$ NMR spectra of these samples suggest that F atoms are included in Cu1-planes in the middle between Cu1 atoms. It is surprising that the magnetic field values for fluorine atoms are similar for samples with different oxygen contents.

### 7.4.3 Mössbauer Study of Y-Ba-Cu-O-$^{129}I$

The samples containing the Mössbauer isotope $^{129}I$ are prepared as described above. Elemental $^{129}I$ enriched to 94% purity ($T_{1/2} = 1.7 \cdot 10^7$ years) was synthesized by solid-phase oxidation of KI by potassium bichromate and purified by sublimation. The synthesized samples were boiled in $CCl_4$ in order to remove any adsorbed iodine. The chemical composition of a typical iodinated crystal was $YBa_2Cu_3O_{6.1}I_{0.96}$ where the content of $^{129}I$ was determined by a method of sulphide leaching described in [7.44]. The superconducting properties and the structure of these iodinated samples are identical to those described above for samples treated with naturally occurring iodine. It has to be noted that no foreign-phase impurities were found by X-ray methods.

Mössbauer spectra of the $5/2^+ \rightarrow 7/2^+$ $^{129}I$ transition with an energy of 27.7 keV were measured in the transmission geometry at 4.2 K [the details of the experimental technique and the spectral analysis are given in [7.31]. Figure 7.12 depicts an example of the spectrum of an iodinated

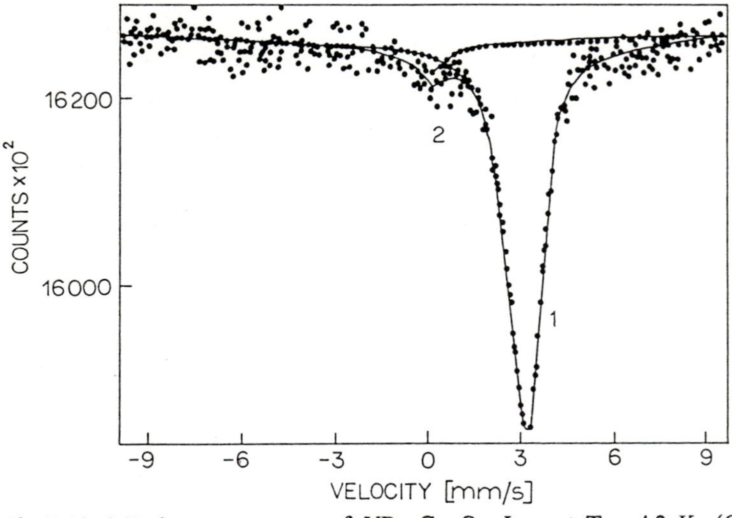


**Fig.7.12.** Mössbauer spectrum of $YBa_2Cu_3O_{6.1}I_{0.96}$ at T = 4.2 K. (Curve *1*: iodide ion; curve *2*: orthorperiodate ion)

sample. The curves are described well by the sum of subspectra corresponding to two chemical forms of iodine [7.29]. These are identified by comparing these isomeric shifts with the known values for $^{129}I$ compounds [7.45]). Thus, the lines correspond to $I^{-1}$ (from N1) and $IO_6^{-5}$ (from N2), that is, with formal valences of -1 and +7. There is no multiplet which would correspond to elementary iodine.

The difference between the experimentally measured values of the isomeric shift $\delta$ = 0.19(1) mm/s in the Y-Ba-Cu-O-I system from $\delta$ = 0.13(1) known for the compound CuI [7.45] is important, too. It allows one to establish the ratio of the iodine in each of the two forms and also to estimate the total iodine content of the sample. Thus, the ratio of form N1 to form N2 is approximately 24:1. Therefore, the $IO_6^{-5}$ ion must be present in the sample in minute quantities relative to $I^{-1}$. The total iodine content can be estimated by the following procedure. The Mössbauer spectra enable one to calculate the effective thickness of the absorber $T_a = n\sigma_0 f$, n being the absorption cross section in the resonance, and f the fraction of $\gamma$-quanta emitted without recoil. The value of f is known from the literature and is variable for iodides within narrow limits: 0.58÷0.64. By setting f = 0.6 (for form N1) [7.46] and f = 0.8 (for form N2) [7.47], one obtains $T_a$ = 1.08 (1) for N1 and $T_a$ = 1.06(1) for N2. In turn, this enables determination of the surface density of the absorber d = 1.05 mg/cm$^2$. Alternatively, direct chemical analysis suggests the chemical formula of the investigated compound, namely $YBa_2Cu_3O_{6.1}I_{0.96}$. That is, the weight content of iodine is 15.8%, which corresponds to a surface density of the absorber of 1.07 mg/cm$^2$. Thus, the agreement between the obtained d value implies that the Mössbauer spectra do yield an accurate value for the total iodine content in the sample. This fact, and also the difference of the magnitude of the isomeric shift of the basic form (N1) - observed experimentally - from isomeric sifts of the simple iodides of Y, Ba or Cu in a single-phase sample (as recorded by X-ray scattering), indicate the appearance of a new chemical form of iodine as a result of a thermal treatment of the $YBa_2Cu_3O_{6.1}$ in iodine vapor.

Measurement of the isomeric shift allows an estimate of the charge $Q_I$ of the iodine in the compound. The result obtained in [7.29], $-0.87 < Q_I < 0$, agrees with the above estimation of the iodine charge derived from the NQR data. Note that a decrease of the iodine charge in Y-Ba-Cu-O-I below -1 implies a smaller value of the iodine radius. This would ease the accommodation of the large iodine ion in the crystal lattice.

### 7.4.4 Mössbauer Studies of Y-Ba-Cu(Fe)-O

Mössbauer spectroscopy using $^{57}Fe$ cores as a probe (Chap.8) is widely employed to obtain information on the chemical and structural features of HTSCs. We, too, applied this method to investigate iodinated samples [7.26]. For this purpose we synthesized samples of Y-Ba-Cu-O ceramics in which copper atoms were partly (to 2, 4 and 10at.%) substituted by $^{57}Fe$ and subjected them to the iodination procedure described in Sect.7.1.3.

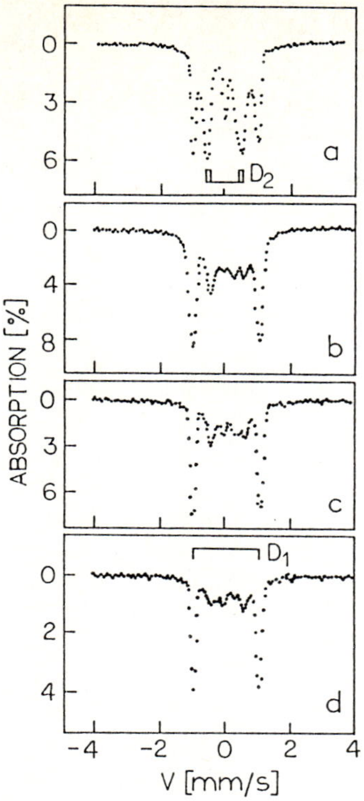


**Fig.7.13a-d.** Mössbauer spectra of Y-Ba-Cu-O ceramics: (**a**) 2 at.% Fe, iodinated; (**b**) 10 at.% Fe, iodinated; (**c**) 10 at.% Fe, oxygen-saturated; (**d**) 10 at.% Fe, prior to iodination. All spectra shown after narrowing of individual components [7.26]

The spectra of the iodinated samples and with varying oxygen and iron contens are displayed in Fig.7.13. The main feature of the fine structure of the orthorhombic ($O_{6.9}$) phases is the tetragonal ($O_{6.1}$) samples only the doublet $D_1$ remains. This is a well documented behavior and is attributed to differences in the environment of the iron cores in these phases. The main result is that after iodination the spectrum does not have the characteristic doublet $D_2$. The same result was obtained in [7.21]. This may indicate the absence of oxygen chains characteristic of the $YBa_2Cu_3O_7$ phase, that is, there is no significant amount of oxygen in Cu1-O type planes after the interaction of the tetragonal samples with iodine. This, in turn, suggests ordered iodine incorporation into the crystal lattice. However, the weakness of the arguments given in [7.21, 26] is the assumption that the copper is substituted by iron mainly in the Cu1 position. This question will be discussed in detail in Chap.8. In particular, the possible difference in the iron-halogen and iron-oxygen interactions compared to the copper-halogen and copper-oxygen interaction in the Y-Ba-Cu-O system is examined. In other words, the electronic structure of iron atoms and the chemical bonds to their nearest environment do not allow one to treat a Mössbauer $^{57}Fe$ ion, incorporated into the Y-Ba-Cu-O lattice, as an independent unperturbing probe.

## 7.5 Substitution Effects in La-Cu-O and Nd-Cu-O. The System Pb-Sr-Cu-O-Cl

In this section we shall briefly discuss a number of studies which seem to lead into promising directions for the in investigation of anion substitutions in HTSCs.

### 7.5.1 Fluorine Doping of $La_2CuO_4$ and $Nd_2CuO_4$

Experiments on anion substitution in the La-Cu-O and Nd-Cu-O systems are especially interesting with regards to, in at least, three aspects. First, the structure of these species is the most simple from among the perovskite-type superconductors. This makes them useful as models to elucidate the mechanism of superconductivity. Second, these compounds are significantly different in structure from Y-Ba-Cu-O, the principal difference being the absence of Cu1-O chains with highly mobile oxygen. Therefore, in these systems the oxygen → halogen substitution presupposes an arrangement of the incorporated atoms in other lattice positions. Finally, these systems have different coupled carries (holes in La-Cu-O and electrons in Nd-Cu-O), which, in principle, makes it possible to observe the correlation of the critical temperature with the type and charge of the incorporated (substituting) atom.

We shall discuss only works concerned with anion substitution in the basic (undoped) systems $La_2CuO_4$ and $Nd_2CuO_4$. Undoped $La_2CuO_4$ is an antiferromagnetic dielectric but it may transform to the superconducting state after annealing at a high oxygen pressure [7.49-52]. This is related to the nonstoichiometricity of oxygen in this compound and, supposedly, to the presence of two structurally similar orthorhombic phases of which one is oxygen-rich ($O_{4+\delta}$). This explains the motivation behind the investigation of partial substitution of oxygen in this system.

*Tissue* et al. [7.53] studied thermally treated $La_2CuO_4$ ceramics (350° C, several seconds) placed in a gas mixture of 10% $F_2$ and 90% $N_2$. A noticeable diamagnetic response ($T_{co} \simeq 35$K) was observed after fluorination. Interestingly, further annealing in air (at 400° C) decreased $T_c$ to 13 K. After fluorination the orthorhombic structure was retained. *Tissue* et al. proposed that fluorine may isomorphically substitute oxygen in certain positions. The fluorine content or other data, supporting the substitution reaction, have not been reported.

Two successful attempts at O → F substitution for the system $Nd_2CuO_4$ have been made. *James* et al. [7.54] and *Koinuma* et al. [7.55] using a method of solid-phase synthesis from CuO, $Nd_2O_3$ and $NdF_3$ with subsequent annealing in nitrogen obtained single-phase samples of $Nd_2CuO_{4-x}F_x$ with $O < x < 0.4$.

The R(T) measurement data showed a superconducting transition at 27 K (for x = 0.3). With x = 0.5, $T_c$ decreased to 10 K, and with x = 0.1 no superconducting transition was seen. For samples with x = 0.2 and 0.4

*James* et al. reported the observation of a 13% Meissner effect. The lattice parameters of Nd-Cu-O-F were close to those for $Nd_2CuO_4$.

*Koinuma* et al. [7.55] investigated the effects of doping thin $Nd_2CuO_4$ films with fluroine. Two procedures were employed: annealing of Nd-Cu-O films in $NF_3$ and deposition of Nd-Cu-O-F films by sputtering in the presence of $NF_3$. Various thermal treatments produced a number of samples showing a superconducting transition with an onset at 27 K [as seen from R(T) measurements]; the resistance, however, did not drop to zero.

### 7.5.2 New Layered Oxychlorides: $Pb_3Sr_3Cu_3O_{8+x}Cl$

Attempts to synthesize perovskite systems in general, and with partial or full substitution of oxygen in particular, are one of the more promising directions in the study of the physics and chemistry of high-temperature superconductors. The pioneering work of *Cava* et al. [7.56] deserves special notice. In 1990 they reported the synthesis and determination of the structure of novel layered oxychlorides: $Pb_3Sr_3Cu_3O_{3+x}Cl$ and $Pb_3Ba_2SrCu_3 \cdot O_8Cl$. These new compounds were obtained by a solid-phase synthesis from the appropriate oxides and $PbCl_2$. *Cava* et al. noted that the samples were single-phase, impregnated with small single crystals.

Without going into the details of interpretaion of a rather complicated structure which, in principle, is similar to that of $Pb_2Sr_2ICu_3O_8$, we point out the presence of triple layers -(Pb/Sr)-Cl-(Pb/Sr)- inherent to this compound. In the opinion of *Cava* et al. chemical doping by a monovalent element or additional oxidation should induce the appearance of superconductivity in this compound.

## 7.6 Conclusion

Numerous experimental facts convincingly demonstrate major changes and the phenomenon of "superconductivity recovery" in the initially dielectric Y-Ba-Cu-O system treated in halogen vapors in the absence of air, $T_c$ being dependent (at least for iodination) on the type of halogen. It would seem that these data may be directly interpreted, noting the presence of vacant positions in Cu1-O planes in $YBa_2Cu_3O_6$, as a result of partial substitution of oxygen by halogen with a simultaneous formation of the orthorhombic phase and charge redistribution, leading to the appearance of charge in Cu2-O planes and the occurrence of HTSC. However, some disagreements, prompted by the use of various experimental methods and the lack (or inconsistency) of direct evidence obtained by structural methods for the incorporation of halogen atoms into the yttrium-barium cuprate lattice, make the interpretation of the observed phenomena more problematic. Thus, in summary, one may consider two possible consequences of the treatment of tetragonal $YBa_2Cu_3O_6$ samples in vapors of halogens or their compounds:

• Halogens react chemically with a solid cuprate: oxygen is released to participate in an additional oxidation of the sample. As a result, a small amount of the orthorhombic phase is produced. The occurrence of this phase of a normal oxygen type is responsible for the observed superconductivity.

• Halogens are incorporated into the crystal lattice of the starting compound. Here, too, there may be two possibilities.

a) Halogen atoms, settling in the appropriate lattice sites and, probably getting ordered, participate in chemical bonds stipulating the appearance of superconducting-current carriers. These positions of halogen atoms in the crystal lattice may be of the Cu1 type or others. (A halogen might also occupy different positions simultaneously, and the situation may be different for different halogens due to differences of their electronic and chemical properties and ion sizes).

b) A halogen atom substitutes for a fraction of an oxygen in $YBa_2Cu_3O_6$ and forces it out into an usual position corresponding to the orthorhombic phases of $YBa_2Cu_3O_7$ or $YBa_2Cu_3O_{6.5}$. In this case the formation of the superconducting electronic subsystem is related, like in all $MeBa_2Cu_3O_{7-\delta}$ (Me: "metal"), to the chemistry of a crystal containing excess oxygen.

The above experimental data suggest that there are serious arguments for each of these models. This complicates an unambiguous interpretation of the observed phenomena, but is also a good stimulus for further efforts. Both the development of new physical methods to investigate substitution effects as well as attempts to obtain more perfect single crystals and films of halogenated Y-Ba-Cu-O must be pursued. The effects of partial substitution in new systems must be understood. Certain measures have already been taken in this direction (Sect.7.5). Every investigation of systems with a deliberate substitution of oxygen will be justified since it will contribute to the general understanding of the HTSC phenomenon and direct the syntheses of new HTSC systems.

## References


7.1 B. Raveau, C. Michel, M. Hervieu, D. Groult: *Crystal Chemistry of High-$T_c$ Superconducting Copper Oxides*, Springer Ser. Mater. Sci., Vol.15 (Springer, Berlin, Heidelberg 1991)

7.2 A.W. Sleight: Science **242**, 1519 (1988)

7.3 S.R. Ovshinsky, R.T. Young, D.D. Allred, G. DeMaggio, G.A. Van der Leeden: Phys. Rev. Lett. **58**, 2579 (1987)

7.4 J.S. Kim, J.S. Swinnea, A. Manthiram, H. Steinfink: Solid State Commun. **66**, 287 (1988)

7.5 P.K. Davis, J.A. Stuart, D. White, C. Lee, P.M. Chaikin, M.J. Naughton, R.C. Yu, R.L. Ehrenhaufer: Solid State Commun **64**, 1441 (1987)

7.5 H.H. Wang, A.M. Kini, H.I. Kao et al.: Inorg. Chem. **27**, 5 (1988)

7.6 T.A. Vanderah, D.L. Decker, D.C. Harris, B.L. Chamberland: Mat. Res. Bull. **24**, 121 (1989)

7.7 Yu. A. Ossipyan, O.V. Zharikov, N.S. Sidorov, V.I. Kulakov, D.N. Mogilyanski-i, R.K. Nikolaev, V. Sh. Shekhtman, O.A. Volegova, I.M. Romanenko: JETP Lett. **48**, 246 (1988)
7.8 J.A. Stuart, P.K. Davies, T.P. Feist: Solid State Ionics **32/33**, 1100 (1989)
7.9 K.M. Cirillo, J.C. Wright, J. Seuntjens, M. Dauemling, D.C. Larbalestier: Solid State Commun. **66**, 1237 (1988)
7.10 S.M. Zahurak, D.M. Murphy, S. Nakahara, W.W. Warren, Jr., D.M. Krol, A.M. Thayer, D.C. Douglass, G.F. Brennert, L.H. Glarum, E.M. Gyorgy, S.M. Fine, M. Greenblatt: Solid State Ionics **32/33**, 1104 (1989)
7.11 N.N. Sauer, E. Garsia, J.A. Martin, R.R. Ryan, P.G. Eller, J.R. Tesmer, C.J. Maggiore: J. Mat. Res. **3**, 813 (1988)
7.12 U.P.K. Rao, A.K. Tyagi, S.J. Patwe, R.M. Iyer, M.D. Sastry, R.M. Kadam, Y. Babu, A.G.I. Dalvi: Solid State Commun. **67**, 385 (1988)
U.P.K. Rao, A.K. Tyagi, S.J. Patwe, R.M. Iyer: Pramana-J. Phys. **31**, L79 (1988)
7.13 C. Perrin, O Pena, M. Sergent, P.S. Christensen, G. Fonteneau, J. Lucas: Physica C **153-155**, 934 (1988)
7.14 J.R. LaGraff, E.C. Behrman, J.A.T. Taylor, E.J. Rotella, J.D. Jorgensen, L.Q. Wang, P.G. Mattocks: Phys. Rev. B **39**, 347 (1989)
7.15 J. Conard, C. Perrin, O. Pena, M. Sergent, G. Fonteneau: Synthetic Metals **34**, 461 (1989)
7.16 C. Perrin, O. Pena, M. Sergent, G. Fonteneau, J. Lucas: Europhys. Lett. **11**, 181 (1990)
7.17 C. Perrin, O. Pena, A. Dinia, M. Sergent, G. Fonteneau, P.S. Christensen, J. Lucas, P. Burlet, J. Rossat-Mignod: Physica C **162-164**, 889 (1989)
7.18 Yu.A. Ossipyan, O.V. Zharikov: Physica C **162-164**, 79 (1989)
7.19 A.G. Klimenko, V.I. Kuznetsov, Ya.Ya. Medikov, A.P. Nemudry, Yu.T. Pavlyukhin, N.G. Khainovsky: Superconductivity 2, 5 (1989) (in Russian)
7.20 A.G. Klimenko, S.M. Ishikaev, Yu.I. Mironov, V.P. Fedin, V.E. Fedorov, M. Sheer: Superconductivity 2, 142 (1989) (in Russian)
7.21 Yu.T. Pavlyukhin, A.P. Nemudry, N.G. Khainovsky, V.V. Boldyrev: Solid State Commun. 72, 107 (1989)
7.22 C. Perrin, O. Pena, M. Sergent: Conf. on high-temperature superconductors, material aspects (Deutsche Gesellschaft für Materialkunde, Oberursel 1990)
7.23 H.B. Radouski, R.S. Glass, D. Back, A.H. Chin, M.J. Fluss, J.Z. Liu, W.D. Mosly, P. Klavins, R.N. Shelton: Preprint (1990)
7.24 P.C.W. Chu: Inside R&D **19**, No.10 (1990)
7.25 Yu.A. Ossipyan, O.V. Zharikov, G.V. Novikov, N.S. Sidorov, V.I. Kulakov, L.V. Sipavina, R.K. Nikolaev, A.M. Gromov: J. Expt. Theor. Phys. Lett. **49**, 73 (1989)
7.26 Yu.A. Ossipyan, O.V. Zharikov, G.V. Novikov, N.S. Sidorov, V.I. Kulakov, L.V. Sipavina, R.K. Nikolaev, A.M. Gromov: Physica C **159**, 137 (1989)
7.27 Yu.A. Ossipyan, O.V. Zharikov, G.Yu. Logvenov, N.S. Sidorov, V.I. Kulakov, I.M. Shmytko, I.K. Bdikin, A.M. Gromov: Physica C **165**, 107 (1990)
7.28 H.B. Radousky, R.S. Glass, P.A. Hahn, M.J. Fluss, R.G. Meisenheimer, B.P. Bonner, C.I. Mertzbacher, E.M. Larson, K.D. Keegan, J.C. O'Brien, J.L. Peng, R.N. Shelton, K.F. McCarty: Phys. Rev. B **41**, 11140 (1990)
7.29 Yu.A. Ossipyan, O.V. Zharikov, A.M. Gromov, V.I. Kulakov, R.K. Nikolaev, N.S. Sidorov, Yu.S. Grushko, Yu.V. Ganzha, M.F. Kovalev, L.I. Molkanov, E.F. Makarov, A.T. Mailybaev: Physica C **171**, 311 (1990)
7.30 R.J. Cava, B. Batlogg, C.H. Chen, E.A. Rietman, S.M. Zahurak, D. Werder: Phys. Rev. B **36**, 5719 (1987)

7.31 R.J. Cava, B. Batlogg, K.M. Rabe, E.A. Rietman, P.K. Gallagher, L.W. Rupp Jr.: Physica C **156**, 523 (1988)
7.32 V.F. Degtyarova, O.V. Zharikov, I.N. Kremenskaya, M.A. Nevedomskaya, R.K. Nikolaev, Yu.A. Ossipyan, A.V. Palnichenko, N.S. Sidorov, V.Sh. Shekhtman: Solid State Commun. **70**, 561 (1989)
7.33 C. Perrin, A. Dinia, O. Pena, M. Sergent, P. Burlet, J. Rossat-Mignod: Solid State Commun. **76**, 401 (1990)
7.34 C. Perrin, O. Pena, M. Sergent, P. Christensen, G. Fonteneau, J. Lucas: Supercond. Sci. Technol. 2, 35 (1989)
7.35 Yu.A. Ossipyan, E.L. Bokhenkov, I.V. Golosovsky, O.V. Zharikov, A.V. Krasheninnikov, S.A. Kuznetsov, R.K. Nikolaev, V.P. Plakhtii, E.G. Ponaytovskii, V.I. Raschupkin, N.S. Sidorov, A.E. Sovestnov, E.I. Fedorova, V.P. Kharchenkov: J. Expt. Theor. Phys. Lett. **49**, 214 (1989)
7.36 A.P. Nemudry, Y.T. Pavlyukhin, N.G. Khainovskii, V.V. Boldyrev: Superconductivity **3**, 1528 (1990) (in Russian)
7.37 O.V. Misochko, Yu.A. Ossipyan, O.V. Zharikov, R.K. Nikolaev, N.S. Sidorov, V.I. Kulakov, A.M. Gromov: Superconductivity **4**, 954 (1991) (in Russian)
7.38 Yu.A. Ossipyan, Yu.S. Greznev, V.L. Matukhin, I.A. Safin, N.S. Sidorov, G.V. Teitelbaum, O.V. Zhariokov: Solid State Commun. **74**, 617 (1990)
7.39 Yu.A. Ossipyan, O.V. Zharikov, V.N. Anashkin, V.L. Matukhin, A.I. Pogoreltsev, I.A. Safin, G.B. Teitelbaum: Z. Naturforsch. **47** A, 21 (1992)
7.40 H. Yasuoka, T. Shimizu, T. Tmai, S. Sasaki, Y. Ueda, K. Kosuge: Hyperfine Interactions, **49**, 167 (1989)
7.41 H. Yasuoka, T. Shimizu, Yu. Ueda, K. Kosuge: J. Phys. Soc. Jpn. **57**, 2659 (1988)
7.42 T. Imai, T. Shimizu, T. Tsuda, H. Yasuoka, T. Takabatake, Y. Nakazawa, M. Ishikawa: J. Phys. Soc. Jpn. **57**, 1771 (1988)
7.43 Y. Hakamichi, K. Kumagai, H. Nakajima, T. Fujita: Physica C **162–164**, 185 (1989)
7.44 Yu. Grushko: USSR Patent No. 417007 (26 October 1973)
7.45 H. de Waard: *Mössbauer Effect Data Index* (Plenum, New York 1975)
7.46 G.J. Kemerink, N. Ruvi, H. de Waard: J. Phys. C **19**, 4897 (1986)
7.47 M. Pasternak, M. Van der Heyden, G. Langouche: J. Chem. Phys. **80**, (1) (1984)
7.48 R.A. Brand, Ch. Sauer, H. Lütgemeier, B. Rupp, W. Zinn: Physica C **156**, 539 (1988)
7.49 J.D. Jorgensen, B. Dabrowskii, Shiyou Pei, D.G. Hinks, L. Soderholm, B. Morosin, J.E. Schirber, E.L. Venturini, D.S. Ginley: Phys. Rev. B **38**, 11337 (1988)
7.50 N.D. Shinn, J.W. Rogers, Jr., E. Schirber, E.L. Venturini, B. Morosin, D.S. Ginley: Synthetic Metals **29**, F709 (1989)
7.51 R. Yoshizaki, H. Sawada, T. Iwazumi: Synthetic Metals **29**, F735 (1989)
7.52 C. Chaillout, S.W. Cheong, Z. Fisk, M.S. Lehmann, M. Marezio, B. Morozin, J.E. Schirber: Physica Scripta T **29**, 97 (1989)
7.53 B.M. Tissue, K.M. Cirillo, J.C. Wright, M. Daeumling, D.C. Larbalestier: Solid State Commun. **65**, 51 (1988)
7.54 A.C.W.P. James, S.M. Zahurak, D.W. Murphy: Nature **338**, 240 (1989)
7.55 H. Koinuma, K. Takeuchi, M. Yoshimoto, K. Hashida, M. Nakabayashi, S. Gonda, T. Hirayama, T. Shiraishi: Jpn. J. Appl. Phys. **29**, L1642 (1990)
7.56 R.J. Cava, P. Bordet, J.J. Capponi, C. Chaillout, J. Chenavas, T. Fournier, E.A. Hewat, J.L. Hodeau, J.P. Levy, M. Marezio, B. Batlogg, L.W. Rupp Jr.: Physica C **167**, 67 (1990)

# 8. Mössbauer Study of Compounds of the Y-Ba-Cu-O System

V. Sedykh

Since the discovery of High-$T_c$ SuperConductors (HTSCs), considerable efforts have been made to understand the effect of impurity ions on the superconducting behavior. Mössbauer spectroscopy has contributed considerably to the understanding of superconductors with impurities. Many papers focus on Mössbauer studies of Fe-doped $YBa_2Cu_3O_{7-x}$ compounds. Although it is universally accepted that CuO planes play a major role in the transport properties of superconducting copper oxides, the mechanism of the substitution of Cu by Fe ions has attracted much attention in the literature. Now it is known that Fe substitutes both Cu1-chain and Cu2-plane sites in the lattice. The Mössbauer method gives valuable information about the local properties around Fe ions, such as site symmetry, local distortion, valence state and magnetic ordering.

## 8.1 Principles of Mössbauer Spectroscopy

The principle of the Mössbauer effect is that the emission and absorption of $\gamma$-rays of atoms fixed in a solid can occur in a recoil-free manner, i.e., withouth energy loss [8.1,2]. Among Mössbauer isotopes $^{57}$Fe has been studied the most. The popularity of this isotope is due to its low transition energy from the first excited nuclear state to the ground state (14keV), the very narrow spectral line with a natural line width $\Gamma$ of about $10^{-9}$ eV, a very high resolution, i.e., the ratio of the line width to the energy of the $\gamma$-rays, $\Gamma/E_\gamma \simeq 10^{-13}$, and finally, the $^{57}$Co source has a rather long lifetime, with a 267-day half-life decaying to the $^{57}$Fe excited nuclear states.

To obtain a Mössbauer spectrum in the usual way there are four components needed: a $\gamma$-ray source, an absorber, a drive system to compensate for the recoil energy loss of the nuclei by motion of the emitting or absorbing nuclei (linear Doppler effect), and a detector with a counting system. The resonance occurs between two identical nuclei, e.g., $^{57}$Fe, in the source and the absorber in their first excited and ground nuclear states.

The Mössbauer spectrum reflects the nature and the strengths of the hyperfine interactions between the Mössbauer nucleus and the surrounding fields. There are three main hyperfine interactions:

- electric monopole interaction resulting in an isomer shift, which measures the s-electron density at the nucleus;
- electric quadrupole interaction resulting in a quadrupole splitting, a measure of the Electric Field Gradient (EFG) at the nucleus; and



Springer Series in Materials Science, Vol. 23
**The Real Structure of High-$T_c$ Superconductors**
Editor: V.Sh. Shekhtman 

- magnetic dipole interaction resulting in the nuclear Zeeman effect, a measure of the internal magnetic field at the nucleus.

The latter two interactions remove the degeneracy of the nuclear levels.

Our discussion will be focused mainly on the electric monopole and quadrupole interactions because mainly the isomer shift and the quadrupole splitting are used in the spectral analysis of superconducting compounds.

### 8.1.1 Isomer Shift

The electron charge density at a nucleus is mainly due to s-electrons of different shells. The isomer shift is proportional to the s-electron density at the nucleus of the emitter or the absorber. The electric interaction of the s-electrons with the atomic nucleus of finite volume results in a shift of the level of the nuclear energy by the very small value $\delta E$ (Fig. 8.1). The s-electron density at the nucleus can be changed by the chemical environment. In a Mössbauer experiment, where an appropriate Doppler velocity is applied to either the source or the absorber in order to get resonance in emitting or absorbing a $\gamma$-ray, the difference of the electrostatic isomer shift between a source (S) and an absorber (A) is observed, which is written in the form [8.2]

$$\delta = C(\delta R/R)[\Psi_A(0)^2 - \Psi_S(0)^2] , \tag{8.1}$$

where C is a constant for a given Mössbauer isotope, $\delta R = R_e - R_g$ is the change in the nuclear radius between the exciting and the ground states, and the term in square brackets is the difference in the electron density between the same isotopes in the absorber and the source, but in different chemical environments. The change in chemical environment can be due to

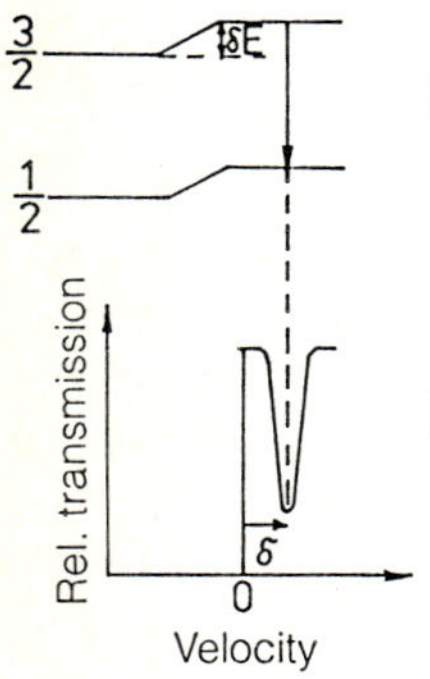


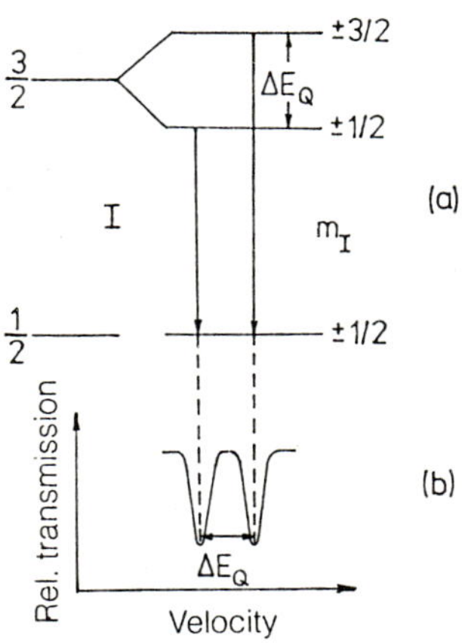


**Fig. 8.1.** (**a**) Isomer shift $\delta E$ of nondegenerate nuclear energy levels; spin I = 3/2 is the excited state, I = 1/2 is the ground state of $^{57}Fe$; (**b**) resultant Mössbauer spectrum (schematic)

**Fig. 8.2.** (**a**) Quadrupole splitting of the nuclear level with spin I = 3/2 in the excited state for $^{57}Fe$; (**b**) resultant Mössbauer spectrum (schematic)

a change of the valence state or the chemical bonding. This means that the Mössbauer isomer shift can yield very useful information about the valence state, the bonding properties and the oxidation state of a Mössbauer atom.

### 8.1.2 Quadrupole Splitting

A consideration of the isomer shift presupposes a spherically symmetric and uniform distribution of nuclear charge. However, there are many cases in which the nuclear charge distribution deviates from spherical symmetry. This results in an interaction of the nuclear electric quadrupole moment with the inhomogeneous Electric Field Gradient (EFG) at the nucleus. The electric quadrupole interaction splits the first excited nuclear state into sublevels, as shown in Fig. 8.2 for $^{57}$Fe. An expression for the quadrupole splitting can be written as [8.2]

$$E_Q = \frac{eQV_{zz}}{4I(2I-1)}[3m_I^2 - I(I+1)]\sqrt{1+\eta^2/3}\ , \tag{8.2}$$

where eQ is the nuclear electric quadrupole moment, $V_{zz}$ is the EFG, I is the nuclear spin quantum number, $m_I$ = I, I-1, ..., -I is the nuclear magnetic spin quantum number, and $\eta$ is the asymmetry parameter defined by

$$\eta = (V_{xx} - V_{yy})/V_{zz}\ . \tag{8.3}$$

As an example, let us consider the electric quadrupole interaction for $^{57}$Fe, where I = 3/2 for the first excited state and I = 1/2 for the ground state (Fig. 8.2). The nuclear ground state is not split, because for $^{57}$Fe Q = 0. The excited state is split into two doubly degenerate sublevels. Assuming, for simplicity, an axially symmetric EFG the expressions for each sublevel can be written as[8.3]

$$E_Q(\pm 3/2) = eQV_{zz}/4 \quad (I = 3/2,\ m_I = \pm 3/3)\ ,$$

$$E_Q(\pm 1/2) = -\ eQV_{zz}/4 \quad (I = 3/2,\ m_I = \pm 1/2)\ . \tag{8.4}$$

In this case the Mössbauer spectrum shows the difference of energy between the two sublevels

$$\Delta E_Q = E_Q(\pm 3/2) - E_Q(\pm 1/2) \tag{8.5}$$

and the distance between two lines of the Mössbauer spectrum corresponds to the energy splitting $\Delta E_Q$.

### 8.1.3 Magnetic Splitting

In magnetic materials the interaction of the nuclear magnetic dipole moment with the magnetic field at the nucleus results in a splitting of the nuclear state with spin I > 0 into (2I+1) sublevels, each of them being charac-

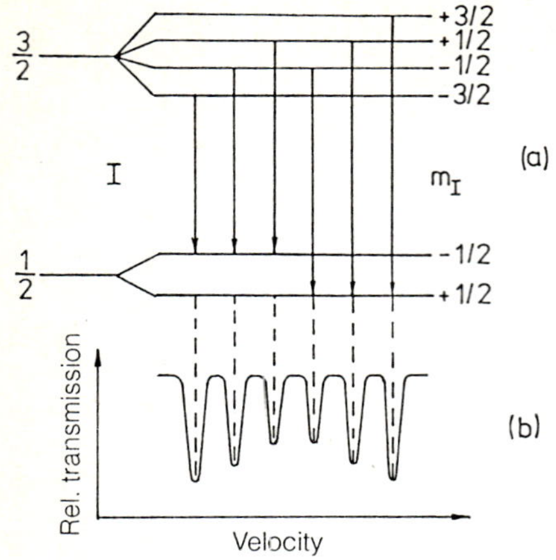


**Fig.8.3.** (a) Magnetic dipole splitting; (b) resultant Mössbauer spectrum (schematic)

terized by the nuclear magnetic spin quantum number $m_I$ = I, I-1, ..., -I (nuclear Zeeman effect). For the $^{57}$Fe Mössbauer isotope the ground state with I = 1/2 splits into two sublevels, and the first excited state with I = 3/2 splits into four sublevels. Taking into account the selection rule $\Delta m = 0,\pm 1$, there are six allowed transitions between the sublevels of the ground and excited states. This corresponds to a six-line pattern of the $^{57}$Fe Mössbauer spectrum, as shown in Fig.8.3. More details about Mössbauer parameters can be found, for example, in [8.1].

Thus, the hyperfine interactions which are the interactions between a nuclear property and an electronic and/or atomic property allow one to get information about the local properties of materials, and about the local environment of a Mössbauer atom. An actual Mössbauer spectrum is a superposition of many simple spectra which each correspond to a certain local environment of the Mössbauer atom. One of the main and most difficult problems is a careful analysis and finding a unique decomposition of a complex Mössbauer spectrum.

## 8.2 Effect of an Fe Impurity on $T_c$ and the Structure of the $YBa_2Cu_3O_{7-x}$ Superconductor

A very important problem of high-$T_c$ superconductivity is to find the mechanism of the occurrence of superconductivity and the role of the Cu1-chain and the Cu2-plane sites. Studies of superconducting compounds with impurities, for instance 3d metal elements (Fe, Co), can help us to unravel some of these problems. It is now known that the superconducting transition temperature $T_c$ decreases when a part of the Cu site is replaced by another 3d metal, but superconductivity is found at low temperatures even with rather significant content of 3d metal [8.4-6]. Therefore Mössbauer measurements of the $^{57}$Fe-doped $YBa_2Cu_3O_{7-x}$ (named 1-2-3) superconducting compounds should be very useful for the investigation of the microscopic properties of the superconducting state, and may give us information about the local symmetry around the Cu1 and the Cu2 sites.

The mechanism causing suppression of $T_c$ with increasing impurity content is not yet understood. Schemes proposed in the literature include disorder in the oxygen distribution, change of local symmetry, and electronic processes such as band filling and electron localization [8.6-8].

It is well known that the crystal structure of the 1-2-3 compound can be either orthorhombic or tetragonal depending on the deficiency or disordering of oxygen [8.9]. There are two distinct sites for the Cu atoms, Cu1-chain and Cu2-plane sites. In the Fe-doped 1-2-3 compounds the Fe atoms occupy both Cu1 and Cu2 sites, although preferentially the Cu1 site for small y [8.10]. $T_c$ decreases monotonically with increasing Fe content up to 12% Fe [8.11]. The transition from orthorhombic to tetragonal structure takes place at the Fe content $y \simeq 0.02$ [8.9-11]. There is no $T_c$ feature in the transition region. In addition, increasing oxygen content is observed with increasing y which reaches $7-x = 7.2$ at $y = 0.5$; superconductivity does not exist for $y > 0.1$ [8.10].

Thus in the Fe-doped 1-2-3 compound there are two types of tetragonal phase: One is induced by removing oxygen in samples quenched from high temperature and is not superconductive. The other tetragonal phase induced by Fe impurities is the bulk superconductivity and is not accompanied by a great number of oxygen deficiencies.

## 8.3 Coordination and Valence of Iron Sites in $YBa_2(Cu_{1-y}Fe_y)_3O_{7-x}$

Despite intense study, at present there is no uncontested information regarding coordination, valence, and amount of Fe locations in $YBa_2 \cdot (Cu_{1-y}Fe_y)_3O_{7-x}$ over the concentration ranges $0.0<y<0.3$, $6.1<7-x<7.3$. There is some discrepancy in the data even among those experiments which dealt with stoichiometrically similar and well characterized samples. This is caused by the extreme sensitivity of the oxygen and iron positions to the thermal history of sample preparation [8.12-14]. However, the situation is not hopeless. Mössbauer results showed that the position of each spectrum component is stable and does not change for different sample preparation and heat treatments [8.12, 14-17]. There is only a change in the relative intensity of the individual components, e.g., the relative iron population at the different sites.

The Mössbauer spectra of the Fe-doped 1-2-3 compounds are rather complicated. To separate them into the correct subspectra it is reasonable to analyze carefully the spectra of the superconducting and nonsuperconducting states, and the kinetics of the spectral line intensity changes after heat treatment. A detailed analysis of the Mössbauer results for the $YBa_2Cu_{2.95} \cdot Fe_{0.05}O_{7-x}$ compound which was prepared by various heat treatment will be presented below [8.12].

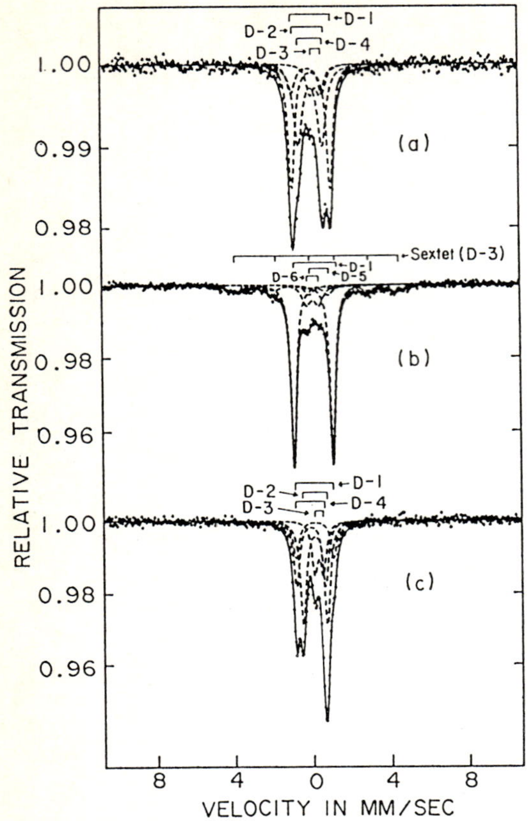


**Fig.8.4a-c.** Typical $^{57}Fe$ Mössbauer spectra at 300 K obtained from $YBa_2Cu_{2.95}Fe_{0.05}O_{7-x}$ prepared by various heat treatments. (**a**) Slowly cooled from 1193 K in air; (**b**) quenched from 1193 K in air into liquid $N_2$;(**c**)oxygen loaded by annealing for 48 hours at 723 K. *Solid line* represents the result from a least-square fitting procedure and the *dashed lines* give the partial components. The velocity scale is relative to bcc Fe at room temperature

Figure 8.4 shows typical $^{57}Fe$ Mössbauer spectra at 300 K obtained from $YBa_2Cu_{2.95}Fe_{0.05}O_{7-x}$. Figure 8.4a presents the spectrum after the sample was slowly cooled from 1193 K in air. The specimen is a superconductor with orthorhombic structure. The spectrum consists of four different quadrupole-split doublets denoted by D-1, D-2, D-3 and D-4 in Fig.8.4a. Figure 8.4b depicts the spectrum obtained just after quenching in air from 1193 K to 77 K. The specimen is a semiconductor with a high-temperature tetragonal structure. After quenching, the D-2 and D-4 components disappear and a magnetically split sextet appears. The sextet is interpreted as the component due to the Fe atoms at the Cu2 sites in the $CuO_2$ plane which shows an antiferromangetic long-range order, as confirmed by a neutron diffraction experiment [8.18]. Above 420 K this magnetic sextet was converted to a paramagnetic double which is identical to the D-3 component in Fig.8.4a. Figure 8.4c displays the spectrum obtained from the specimen which was annealed for 48 hours at 723 K in flowing oxygen gas after quenching from 1193 K. After annealing in oxygen at 723 K the magnetic component disappears and the D-2, D-3 and D-4 components reappear. The orthorhombic structure is recovered after the oxygen loading and the intense D-2 and D-4 components were observed, as shown in Fig.8.4c. From these experimental findings, the D-3 component is surely attributed to the Fe atoms in the Cu2-plane sites and the other components are attributed to the Fe atoms in the Cu1-chain sites having different nearest neighbor configurations of the oxygen atoms. This is because diffusion of Fe atoms can hardly change the sites from a plane to a chain at low temperature. The D-2 and D-4 components are due to the Fe atom having larger numbers of oxygen atoms than those of D-1, D-5 and D-6, because larger intensities of the former were observed when the oxygen concentration was increased by oxygen loading.

To pursue the transformation kinetics of the superconducting-to-nonsuperconducting state the sample was vacuum annealed at different temperatures and for different time periods. The typical Mössbauer spectra obtained in the course of vacuum annealing are exhibited in Fig.8.5a-e.

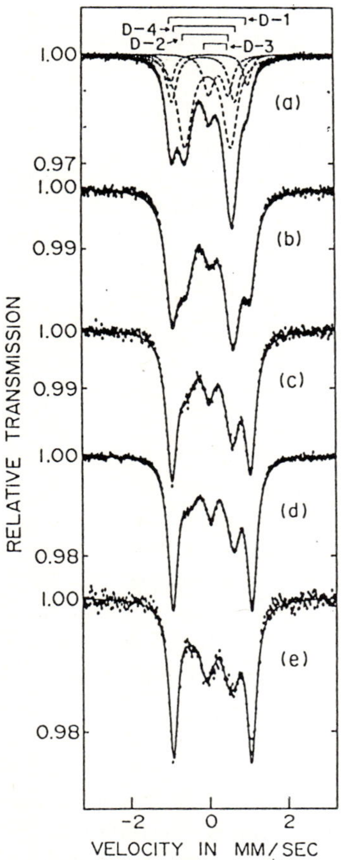


**Fig.8.5a–e.** Typical $^{57}Fe$ Mössbauer spectra at 300 K obtained in the course of vacuum annealing from 450 to 700 K in 50 K steps. (**a**) Original spectrum after oxygen loading; after vacuum-annealing (**b**) for 3 hours at 500 K, (**c**) for 1 hour at 550 K, (**d**) for 3 hours at 550 K, (**e**) for 1 hour at 650 K. Solid lines present a least-square fit. Only in (**a**) the resolved partial components are shown by *dashed lines*. The velocity scale is relative to bcc Fe at room temperature

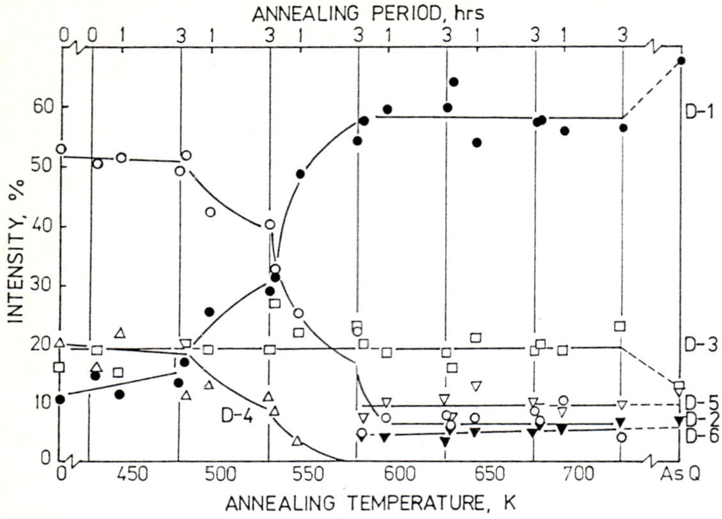


**Fig.8.6.** Intensity changes of the spectral components during vacuum annealing with annealing temperature and period. *Solid curves* are drawn to guide the eye. The notations are: • D-1, ○ D-2, □ D-3, △ D-4, ▽ D-5 and ▼ D-6. The notation As Q (*lower right*) shows the results immediately after quenching from 1193 K in air

Figure 8.5a shows the spectrum obtained from the original speciment immediately after annealing in $O_2$ which is identical to Fig.8.4c except in a narrower velocity range. The solid line gives the results from a least-square fit using a thin-foil approximation and the dashed curves represent the resolved partial components from D-1 to D-4. The intensity changes of the components as functions of annealing temperature and the period in vacuum are illustrated in Fig.8.6. The intensity of the D-4 component begins to fall at 500 K and nearly disappears after annealing for 3 hours at 550 K. The D-2 component also shows a remarkable decrease at 500 K and becomes less than 10% of the total intensity after annealing above 600 K. The D-1 component increases with the decrease of the D-2 and D-4 components and saturates above 600 K. The D-3 component remains unchanged during all the vacuum annealing, strongly supporting its assignment to Fe in the Cu2-plane sites. The D-5 and D-6 components appear after annealing above 600 K and their intensities do not depend on annealing temperature and period. It is suggested that these components arise from rather stable configurations of oxygen around Fe. The D-5 and D-6 components are also observed just after quenching from 1193 K and are attributed to Fe at the Cu1- chain sites in oxygen deficient compounds, as is the D-1 component.

These results suggest that oxygen desorption occurs appreciably during vacuum annealing at 500 K and saturates after annealing for 3 hours at 550 K since the spectrum did not show any noticeable change. This means that the change in the nearest-neighbor configuration of oxygen around the Fe atoms associated with the oxygen desorption is almost completed by anneal-

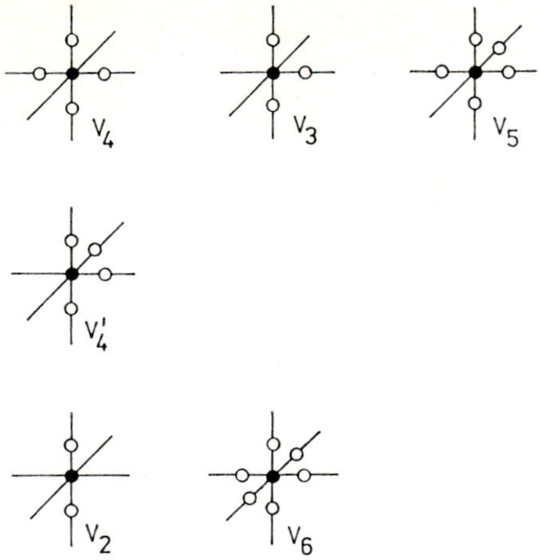


**Fig.8.7.** Six different (idealized) coordinations which may be adopted at an Fe-substituted Cu-chain site

ing at 600 K. After the final annealing at 700 K, the specimen was again reannealed at 723 K for 48 hours in flowing oxygen gas. The resultant Mössbauer spectrum at 300 K showed full recovery of the original one, suggesting that the structural changes occurring under vacuum annealing are reversible, when oxygen is loaded.

Many attempts have been made to determine the Fe configurations with their nearest oxygen neighbors at the Cu1 sites. Analyzing carefully and critically different Mössbauer results and EXAFS measurements [8.19] *Lines* and *Eibschutz* suggested the most debated interpretaion [8.20]. There are six possible configurations for the Cu1 iron sites with their nearest oxygen neighbors, viz. $V_2$, $V_3$, $V_4$, $V_4'$, $V_5$ and $V_6$ (Fig.8.7). Actually, the real configurations may be somewhat different from the idealized configurations in the figure, but it is assumed that they can be identified. The $V_2$ and $V_6$ configurations were shown to be relatively unfavored and can be neglected in a first approximation [8.20]. The analysis revealed that the Fe atoms enter the Cu1 sites with predominantly three-, four-, and five-fold coordinations. The D-1 component can thus be attributed to the $V_3$ coordination with oxygen deficiency, D-2 to the $V_5$ coordination with an oxygen-rich site and D-4 to a square planar coordination $V_4$ or, more likely, to the sites placed along domain boundaries in the distorted tetrahedral configuration $V_4'$.

$^{57}$Fe Mössbauer parameters obtained at 300 K for $YBa_2Cu_{2.95} \cdot Fe_{0.05}O_{7-y}$ are tabulated in Table 8.1. The isomer shift value of D-3 (iron in the Cu2 site) corresponds to the high-spin $Fe^{3+}$ state (S = 5/2). A great deal of controversy surrounds the question of valence and spin quantum state for the Fe atoms in the Cu1 sites. Isomer shift values for D-1 and D-2 are not typical for high-spin $Fe^{3+}$. The isomer shift value of D-4 is negative, showing the largest charge density at the nucleus compared with the other components. A survey of the literature data leads to the conclusion [8.20] that the D-2 component corresponds almost certainly to $V_5$, $Fe^{3+}$ (S = 3/2); D-1 probably to $V_3$, $Fe^{2+}$ (S = 1); and D-4 most likely to $V_4'$, $Fe^{4+}$

**Table 8.1.** $^{57}Fe$ Mössbauer parameters at room temperature obtained for $YBa_2Cu_{2.95}Fe_{0.05}O_{7-x}$. ($\delta$: isomer shift relative to bcc Fe at room temperature [mm/s], $\Delta$: quadrupole splitting [mm/s]). (a) slow cooling in air from 1193 K, (b) quenched from 1193 K, (c) oxygen loading for 48 h at 723 K, and (d) vacuum annealing for 3 h at 700 K

| | (a) | | (b) | | (c) | | (d) | |
|---|---|---|---|---|---|---|---|---|
| | $\delta$ | $\Delta$ | $\delta$ | $\Delta$ | $\delta$ | $\Delta$ | $\delta$ | $\Delta$ |
| D-1 | 0.03 | 1.97 | 0.02 | 2.00 | 0.03 | 1.96 | 0.04 | 1.96 |
| D-2 | 0.00 | 1.22 | - | - | -0.01 | 1.17 | -0.01 | 1.17 |
| D-3 | 0.27 | 0.52 | - | - | 0.27 | 0.50 | 0.26 | 0.50 |
| D-4 | -0.16 | 1.60 | - | - | -0.12 | 1.60 | - | - |
| D-5 | - | - | 0.02 | 0.63 | - | - | 0.01 | 0.63 |
| D-6 | - | - | 0.32 | 0.98 | - | - | 0.32 | 0.97 |
| Sextet[a] | - | - | 0.28 | 0.24[b] | | | | |

[a] magnetic hyperfine field of 256 kOe

[b] $-\frac{1}{2}\Delta(3\cos^2\theta - 1)$ [8.12]

(S = 2), although $V_4$ and $V'_4$ $Fe^{3+}$ (S = 3/2) have not been ruled out completely. However, since none of these valences and configurations were satisfactorily characterized in the literature, the problem cannot be completely closed.

## 8.4 Magnetic Ordering and Superconductivity in $YBa_2(Cu_{1-y}Fe_y)_3O_{7-x}$

Now it is well accepted that superconductivity depends primarily on the Cu2-$O_2$ planes in all classes of the high-$T_c$ superconducting copper oxides. Much effort has been expended to understand the electronic and magnetic structure of these oxide compounds.

The two-dimensional antiferromagnetism in the $YBa_2Cu_3O_{7-x}$ compounds with x > 0.6 was first found by muon-spin-rotation experiments [8.21] and from neutron diffraction results [8. 18]. It was shown that the antiferromangetism is established in the Cu2-$O_2$ planes and that these planes are correlated along the *c*-axis. To confirm the results of [8.18,21] Mössbauer studies of the Fe-doped 1-2-3 compounds may be useful.

Mössbauer spectra of the $YBa_2(Cu_{1-y}Fe_y)_3O_{7-x}$ (y ≥ 0.06) semiconducting compounds quenched from high temperature show, besides a paramagnetic quadrupole doublet, a small magnetic sextet caused by long-range antiferromangetic ordering with the Néel temperature $T_N$ = 420 K [8.6, 12,13,22-24]. With decreasing temperature the main paramnagnetic part of the spectrum corresponding to Fe in the Cu1-chain sites shows a broad complex feature with a magnetic field value much less than that for the Cu2 sites. The freezing temperature for the magnetic ordering of Fe at the Cu1 sites strongly depends on the Fe concentration, while the Néel temperature of Fe at the Cu2-plane sites does not depend on the Fe content. The temperature dependence of the hyperfine field of Fe at the Cu2 sites follows a $T^{3/2}$ law. For Fe at the Cu1 sites the temperature dependence of the hyperfine field deviates remarkably from a Brillouin-like curve [8.24]. This suggests that the exchange field at the Cu1 sites is much weaker than at the Cu2 sites. Thus, in the nonsuperconducting 1-2-3 compound the Fe atoms at the Cu sites show two different magnetic behaviors: the Cu2-plane site exhibits long-range antiferromangetic order and the magnetic interactions at the Cu1-chain sites are very weak.

The oxygen loading depresses the long-range magnetic order. In the superconducting state, instead of a disappearing magnetic doublet, the D-3 component appears corresponding to Fe at the Cu2-plane site (Fig.8.4). This site is paramagnetic until magnetic freezing of Fe at the Cu1-chain sites occurs. The magnetic freezing temperature for the Cu1 sites strongly depends on the Fe concentration. In $YBa_2(Cu_{1-y}Fe_y)_3O_{7-x}$ superconducting compounds the magnetic ordering seen at liquid-helium temperature disappears as the Fe concentration is made lower than 0.03 [8.13]. It has been suggested that the magnetic ordering is due to the Fe-Fe interaction in the Cu1-plane which becomes much weaker at low Fe concentration. In the case of 8% Fe content the magnetic ordering appears at 15 K and $T_c$ is 34 K [8.24]. Coexistence of magnetic ordering and superconductivity suggests that a magnetic order is formed among the Cu and Fe atoms at the Cu1-chain sites while the Fe concentration at the Cu2-plane sites is not enough to destroy the superconductivity. A comparison of the spectra for the 8% Fe sample at 0.1 K and 4.2 K revealed that the degree of broadening is almost the same. If the broadening is caused by a magnetic relaxation, the line width should become smaller with decreasing temperature. It is more likely that the changes are caused by a random orientation of the magnetic moments like in spin-glasses [8.24].

## 8.5 Fe-Doped $YBa_4Cu_3O_{8.5+x}$

The discovery of high-$T_c$ superconductivity in the 1-2-3 compounds has also stimulated considerable effort to identify the different compounds and phases in the $Y_2O$-BaO-CuO system [8.25,26]. For one of them, $YBa_4 \cdot Cu_3O_{8.5+x}$ (named 1-4-3), the positions and occupation numbers of especially the oxygen ions were determined by neutron powder-diffraction measurements [8.26]. In contrast to the 1-2-3 compounds, in the structure

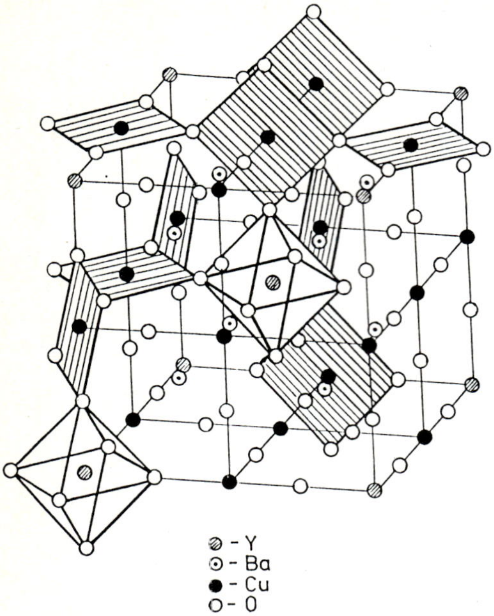


**Fig.8.8.** Unit cell of $YBa_4Cu_3O_{8.5+x}$

of the semiconducting 1-4-3 compound there are only Cu-chain sites along all three crystallographic axes.

The X-ray diffraction pattern of the as-prepared 1-4-3 compound presents a cubic unit cell [8.26, 27]. This cell, as shown in Fig.8.8, is built up of 8 perovskite blocks with the Ba atoms at the centers. The Y atoms are placed both at the center and at the corners of the unit cell and are octahedrally coordinated by O1 and O2 oxygen atoms. The Cu atoms are located on the remaining cation sites of the 8 perovskite blocks. The remaining oxygen atoms are placed between adjacent Cu atoms on the O3 and O4 oxygen positions.

In the 1-4-3-$O_9$ compound about 80% of the O3 sites and 20% of the O4 sites are occupied by oxygen atoms. In the 1-4-3-$O_{8.5}$ compound the oxygen atoms occupy about 98% of the O3 sites and 2% of the O4 sites [8.26]. The crystal structure of 1-4-3 implies a large degree of disorder of the vacancies at the O3 and the oxygen at the O4 sites. The Cu atoms are basically four-fold coordinated by the oxygen atoms. The Cu-O squares are tilted 90° with respect to adjacent squares and form the chains through the O3 oxygen atoms linking the copper atoms. A partial occupancy of the O3 sites means that the chains are not infinite but coordinated in isolated chain fragments. The increasing occupancy of the O3 sites in 1-4-3-$O_{8.5}$ increases the order in the chains. It should again be noted that in the 1-4-3 structure there are only Cu-chain sites, contrary to the structure of the 1-2-3 compound where the Cu atoms occupy two different positions, Cu2-planes and Cu1-chains sites.

To understand the microscopic change in the 1-4-3 structure induced by the rearrangement and removal of the oxygen atoms Mössbauer meas-

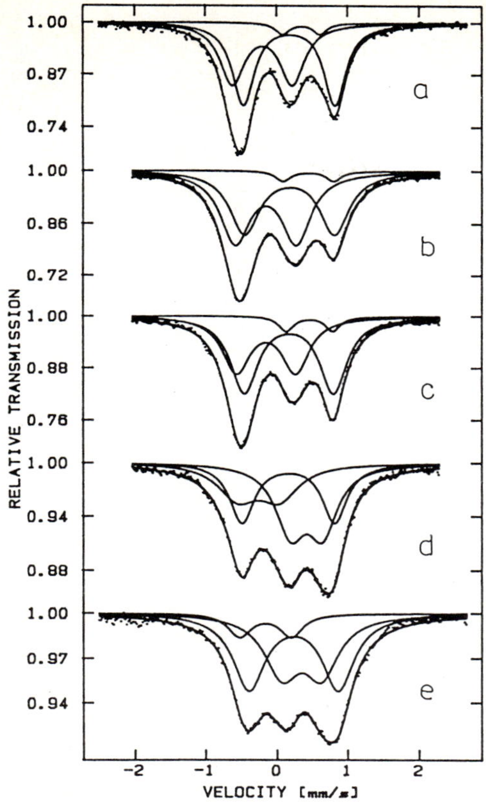


**Fig.8.9a–e.** Typical $^{57}Fe$ Mössbauer spectra of Fe-doped $YBa_4Cu_3O_{8.5+x}$ at room temperature. (**a**) Spectrum of as-prepared sample. After vacuum-annealing for 1 hour at (**b**) 300° C, (**c**) 350° C, (**d**) 400° C, (**e**) 700° C. Velocity scale is relative to bcc Fe at room temperature

urements were carried out on the $YBa_4Cu_{2.95}Fe_{0.05}O_{8.5+x}$ sample annealed at different temperatures in vacuum [8.27]. It was assumed that the Fe atoms also occupy Cu sites as in the Fe-doped 1-2-3 compound [8.12].

Figure 8.9 shows typical $^{57}Fe$ Mössbauer spectra of the studied specimen obtained at room temperature after the heat treatment in vacuum [8.27]. The most remarkable changes in the spectra are observed on the sample annealed at 400° C. All spectra were decomposed into three different quadrupole split doublets A, B and C. Figure 8.9a depicts the spectrum of the as-prepared sample. Mössbauer parameters obtained at room temperature are tabulated in Table 8.2. The isomer-shift values of the B and C components are typical for the formal high-spin $Fe^{3+}$ valence state. The isomer-shift value of the A component is negative and suggests the possibility of the formal $Fe^{4+}$ valence state. The intensity changes of the doublet lines as a function of the vacuum annealing temperature are shown in Fig.8.10. The A component may be attributed to the Fe atoms in a rich oxygen environment and the C component to the Fe atoms in a deficient oxygen environment.

Using the known data for the 1-2-3 compound [8.12], the present results, and the results of [8.20], we can dare to define the configurations of

**Table 8.2.** [57]Fe Mössbauer parameters for three quadrupole split doublets A, B, C (Fig.8.10) at room temperature obtained from vacuum annealed Fe-doped $YBa_4Cu_3 \cdot O_{8.5+x}$. ($\delta$: isomer shift relative to bcc Fe at room temperature [mm/s], $\Delta$: quadrupole splitting [mm/s])

| Annealing Temperature [°C] | A | | B | | C | |
|---|---|---|---|---|---|---|
| | $\delta$ | $\Delta$ | $\delta$ | $\Delta$ | $\delta$ | $\Delta$ |
| as-prepared | -0.199 | 0.837 | 0.347 | 0.530 | 0.181 | 1.287 |
| 300 | -0.152 | 0.853 | 0.445 | 0.718 | 0.183 | 1.264 |
| 350 | -0.148 | 0.827 | 0.462 | 0.658 | 0.175 | 1.254 |
| 400 | -0.255 | 0.579 | 0.422 | 0.456 | 0.168 | 1.305 |
| 700 | -0.160 | 0.724 | 0.351 | 0.561 | 0.241 | 1.259 |

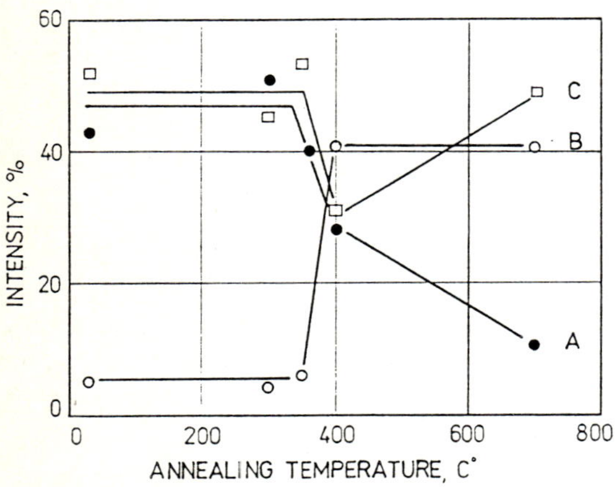


**Fig.8.10.** Intensity changes of the spectral components during vacuum annealing with changing annealing temperature. *Solid curves* only represent a guide for the eye

the Fe sites and their nearest-neighbor oxygen sites. In the first approximation, the $V_2$ and $V_6$ configurations (Fig.8.7) can be eliminated, as argued in Sect.8.3. Since a detailed discussion of the oxygen configurqations associated with the D-5 and D-6 components is not yet possible and because the intensity of these components is smaller than 10% of the total intensity, D-5 and D-6 have not been discussed here, for simplicity. This means that the Fe atoms enter the Cu sites with primarily three-, four-, and five-fold coordinations. The $V_3$-sites dominate in the most highly oxygen deficient samples while the $V_5$ sites are the most preferred in rich oxygenated samples. The four-fold coordination may be represented as a square-planar chain configuration $V_4$ or - along domain boundaries - in the distorted $V_4'$ configuration. The A, B and C components can be attributed to $V_5$, $V_4$ or $V_4'$ and $V_3$ coordinations, respectively. The behavior of the B component allows us to suppose that the transition from the x = 0.5 to the x = 0.0 state arises by a jump at 400° C and that the number of the oxygen-ordering chain sites increases. However, not all of the Fe atoms can achieve the $V_4$ coordination and it is likely that oxygen vacancies occur at the boundaries separating ordered regions, with $V_3$ and $V_5$ coordinated iron located at these boundaries.

The effect of Fe impurities and of the type of site, Cu2-plane or Cu1-chain, on the superconductive transition in 1-2-3 compounds has not been studied in sufficient detail. Recently obtained results for a specially prepared Fe-doped 1-2-3 compound showed that it is possible to obtain a lower ratio of Fe in the Cu1 versus the Cu2 sites compared with conventionally prepared materials, with orthorhombic structure and increased superconductive transition temperature $T_c$ at rather high Fe content (y = 0.1) [8.28]. The increasing Fe occupancy at the Cu2 sites indicated that Fe at Cu2 is not detrimental to superconductivity and may actually play a role in charge fluctuations. On the other hand, because the Fe-doped 1-4-3 compound indicated the presence of Cu-chain sites only and did not show superconductivity, it may be suggested that the Cu2-plane sites play more than a minor role in the superconducting behavior.

## References


8.1 V.S. Shpinel: *γ-Ray Resonance in Crystals* (Nauka, Moscow 1969)

8.2 U. Gonser: From a strange effect to Mössbauer spectroscopy, in *Mössbauer Spectroscopy*, ed. by U. Gonser, Topics Appl. Phys., Vol.5 (Springer, Berlin, Heidelberg 1975) pp.1-51

8.3 P. Gutlich: Mössbauer spectroscopy in chemistry, in *Mössbauer Spectroscopy*, ed. by U. Gonser, Topics Appl. Phys., Vol.5 (Springer, Berlin, Heidelberg 1975) pp.53-96

8.4 G. Xiao, F.H. Streitz, A. Gavrin, Y.W. Du, C.L. Chien: Phys. Rev. B **35**, 8782 (1987)

8.5 Y. Maeno, M. Kato, Y. Aoki, T. Nojima, T. Fujita: Physica B **148**, 357-359 (1987)

8.6 K. Westerholt, H.J. Wuller, H. Bach, P. Stauche: Phys. Rev. B **39**, 680-689 (1989)

8.7 A. Nath, Z. Homonnay: Physica C **161**, 205-208 (1989)

8.8 F. Bridges, J.B. Boyce, T. Claeson, T.H. Gebelle, J.M. Tarascon: Phys. Rev. B **39**, 603-617 (1989)

8.9 Y. Maeno, M. Kato, Y. Aoki, T. Fujita: Jpn. J. Appl. Phys. **26**, L1982-L1984 (1987)

8.10 J.M. Tarascon, P. Barboux, P.F. Miceli, L.H. Greene, G.W. Hull, M. Eibschutz, S.A. Sunshine: Phys. Rev. B **37**, 7458 (1988)

8.11 J. Jing, J. Bieg, H. Engelmann, Y. Hsia, U. Gonser: Solid State Commun. **66**, 727-730 (1988)

8.12 V. Sedykh, S. Nasu, F.E. Fujita: Solid State Commun. **67**, 1063-1067 (1988)

8.13 Z.Q. Qiu, Y.W. Du, H. Tang, J.C. Walker: J. Magn. Magn. Mat. **78**, 359-363 (1989)
8.14 E. Baggio-Saitovitch, I.S. Azevedo, R.B. Scorzelli, H. Saitovitch, S.F. Da Cunha, A.P. Guimaraes, P.R. Silva, A.Y. Takeuchi: Phys. Rev. B **37**, 7967 (1988)
8.15 S. Nasu, H. Kitagawa, Y. Oda, T. Kohara, T. Shinjo, K. Asayama, F.E. Fujita: Physica B **148**, 484-487 (1987)
8.16 L. Bottyan, B. Molnar, D.L. Nagy, I.S. Szucs, J. Toth, J. Dengler, G. Ritter, J. Schober: Phys. Rev. B **38**, 373-381 (1988)
8.17 C. Blue, K. Elgaid, I. Zitkovsky, P. Boolchand, D. McDaniel, W.C.H. Joiner, J. Oostens, W. Huff: Phys. Rev. B **37**, 5905 (1988)
8.18 J.M. Tranquada, D.E. Cox, W. Kunnmann, H. Moudden, G. Shirane, M. Suenaga, P. Zolliker, D. Vaknin, S.K. Sinha, M.S. Alvarez, A.J. Jacobson, D.C. Johnston: Phys. Rev. Lett. **60**, 156 (1988)
8.19 K. Donnelly, J.M.D. Coey, S. Tomlinson, J.M. Greneche: Physica C **156**, 579 (1988)
8.20 M.E. Lines, M. Eibschutz: Physica C **166**, 235-247 (1990)
8.21 N. Nishida, H. Miyatake, D. Shimada, S. Okuma, M. Ishikawa, T. Takabatake, Y. Nakawawa, Y. Kuno, R. Keitel, J.H. Brewer, T.M. Riseman, D.L. Williams, Y. Watanabe, T. Yamazaki, K. Nishiyama, K. Nagamine, E.J. Ansaldo, E. Torikai: Jpn. J. Appl. Phys. **26**, L1856 (1987)
8.22 Q.A. Pankhurst, A.H. Morrish, M. Raudsepp, X.Z. Zhou: J. Phys. C **21**, L7-14 (1988)
8.23 E.R. Bauminger, D. Edery, I. Felner, M. Kowit, Y. Lehavi, I. Nowik: Hyperfine Inter. **47-48**, 560-562 (1989)
8.24 T. Shinjo, S. Nasu: Mössbauer studies of high-$T_c$ oxides, in *Mechanisms of High Temperature Superconductivity*, ed. by H. Kamimura, A. Oshiyama, Springer Ser. Mater. Sci., Vol.11 (Springer, Berlin, Heidelberg 1989) pp.166-175
8.25 D.M. de Leeuw, C.A.H.A. Mutsaers, R.A. Steeman, E. Frikkee, H.W. Zadbergen: Physica C **158**, 391-396 (1989)
8.26 D.M. de Leeuw, C.A.H.A. Mutsaers, C. Langereis, H. Smoorenburg, P.J. Rommers: Physica C **152**, 39-49 (1988)
8.27 V Sedykh, J. Jiang, G.K. Strukova, U. Gonser, F.E. Fujita: Appl. Phys. A **53**, 426-429 (1991)
8.28 M.G. Smith: Phys. Rev. B **42**, 4202-4207 (1990)

# Subject Index

Printing: Saladruck, Berlin
Binding: Buchbinderei Lüderitz & Bauer, Berlin

# Springer-Verlag and the Environment

We at Springer-Verlag firmly believe that an international science publisher has a special obligation to the environment, and our corporate policies consistently reflect this conviction.

We also expect our business partners – paper mills, printers, packaging manufacturers, etc. – to commit themselves to using environmentally friendly materials and production processes.

The paper in this book is made from low- or no-chlorine pulp and is acid free, in conformance with international standards for paper permanency.